Ferroelectrics

Advances in fundamental studies and emerging applications

Online at: https://doi.org/10.1088/978-0-7503-3975-9

Ferroelectrics

Advances in fundamental studies and emerging applications

Edited by
Yang Bai
Microelectronics Research Unit, Faculty of Information Technology and Electrical Engineering, University of Oulu, FI-90570 Oulu, Finland

Ilya Grinberg
Department of Chemistry, Bar-Ilan University, Ramat Gan 529002, Israel

IOP Publishing, Bristol, UK

ISBN 978-0-7503-3975-9 (ebook)
ISBN 978-0-7503-3973-5 (print)
ISBN 978-0-7503-3976-6 (myPrint)
ISBN 978-0-7503-3974-2 (mobi)

DOI 10.1088/978-0-7503-3975-9

Version: 20221201

IOP ebooks

British Library Cataloguing-in-Publication Data: A catalogue record for this book is available from the British Library.

Published by IOP Publishing, wholly owned by The Institute of Physics, London

IOP Publishing, No.2 The Distillery, Glassfields, Avon Street, Bristol, BS2 0GR, UK

US Office: IOP Publishing, Inc., 190 North Independence Mall West, Suite 601, Philadelphia, PA 19106, USA

Contents

Preface

Ferroelectricity was discovered about a century ago. As an electric cousin of the widely known ferromagnetism, ferroelectric materials have shown no fewer applications, from memories to sensors, actuators and transducers, and from military devices to devices in industrial and medical sectors. Because of their structural and functional flexibility and complexity, the fundamental science and applications of ferroelectrics are evolving continuously. Therefore, it is important to provide up-to-date summaries of new discoveries and fundamental progress in understanding this unique physical phenomenon and material family.

In the recent decade, research on ferroelectrics has been rapidly developed in terms of both fundamental and application-oriented studies. This book covers topics of both fundamental and experimental interest and will stress the most significant advances in the past 10 years. In fundamental studies, first-principles calculations based on the density functional theory have been widely used to explore the relationship between ferroelectricity and band structures as well as collective properties such as phase transitions and coercive field. Machine learning has been introduced to predict material properties and thus to speed up materials development. In addition, ferromagnetic models have been adapted to ferroelectrics to predict material property changes with external boundary conditions. In application-oriented studies, advanced ferroelectric ceramics made from band gap engineered compositions have been pioneered for emerging multi-sensing and energy harvesting applications. The domains, domain walls and interfaces in ferroelectrics have been further investigated for use in nanoelectronics. Some of the recently intensively researched halide perovskites for photovoltaic applications have also been proved to be able to exhibit ferroelectric properties. This book collects chapters thoroughly reviewing and summarizing the above recently obtained knowledge through research in the field.

Ferroelectricity is a phenomenon that is simple to describe but is extremely complex to study and understand. This book has its own specific focus on aspects mentioned above, supplementing the gaps in existing books. On the other hand, this book will link the fundamental aspects of ferroelectrics examined in theoretical modelling to the issues encountered in practical use such as microstructure and emerging applications. Such a combination is lacking in the literature but is increasingly needed by ferroelectric researchers who must gain and use knowledge obtained in both fundamental and applied studies. As the theories evolve and computational power increases, experimental studies can no longer solely rely on empirical trial and error. Similar to many other classic fields, the advance from separate fundamental and experimental studies forward to simulation-guided experimental validations will also mark a maturity of ferroelectric research. Readers, potentially materials researchers and electronic/device engineers, will benefit from the book by obtaining the latest verified research results and conclusions reached by well-known, pioneering and leading scientists for each of the above topics. In particular, young and early-career researchers will be inspired by the book to create new ideas to further develop the field in future.

Acknowledgements

Y B acknowledges the supports of University of Oulu. I G acknowledges the support by Israel Science Foundation (Grant No. 1479/21) and the support of the Army Research Laboratory (Grant No. W911NF-19-2-0119).

Editor biographies

Dr Yang Bai

Dr Yang Bai is currently a Docent (Adjunct Professor) and a tenure-track Assistant Professor in Microelectronics Research Unit, Faculty of Information Technology and Electrical Engineering, University of Oulu, Finland. He specializes in research and teaching of ferroelectric and piezoelectric materials and devices, especially in the topics of sensing, energy harvesting and photoferroelectrics. He obtained his PhD degree in 2015 from University of Birmingham, UK. He was awarded the prestigious EU's Marie Sklodowska-Curie Individual Fellowship in 2016 and the European Research Council (ERC) Starting Grant in 2021. He is a lifetime member of the American Ceramic Society (ACerS) and serves as a committee member in the Energy Materials and Systems Division (EMSD) of ACerS. His research has been published in over 40 peer-reviewed journal articles including those in multiple high-impact journals, and he has been granted five patents.

Dr Ilya Grinberg

Dr Ilya Grinberg is currently an Associate Professor at the Department of Chemistry, Bar-Ilan University, Ramat Gan, Israel. His research focuses on the theoretical modeling of complex oxides using first-principles calculations, molecular dynamics simulations and analytical modeling. He obtained his PhD degree in 2002 from University of Pennsylvania, Philadelphia, USA. He is a co-author of more than 80 research publications, including in multiple high-impact journals and has been granted several patents.

List of contributors

Yang Bai
Microelectronics Research Unit, University of Oulu, Finland

Benjamin Ducharne
ELyTMaX IRL 3757, CNRS, INSA Lyon, Centrale Lyon, Université Claude Bernard Lyon 1, Tohoku University, Sendai, Japan

Ilya Grinberg
Depeartment of Chemistry, Bar-Ilan University, Ramat Gan, Israel

Yihao Hu
Key Laboratory for Quantum Materials of Zhejiang Province, Department of Physics, School of Science, Westlake University, Hangzhou, Zhejiang, China

Jiawei Huang
Key Laboratory for Quantum Materials of Zhejiang Province, Department of Physics, School of Science, Westlake University, Hangzhou, Zhejiang, China

Shi Liu
Key Laboratory for Quantum Materials of Zhejiang Province, Department of Physics, School of Science, Westlake University, Hangzhou, Zhejiang, China

Zhuang Qian
Key Laboratory for Quantum Materials of Zhejiang Province, Department of Physics, School of Science, Westlake University, Hangzhou, Zhejiang, China

Holger Röhm
Light Technology Institute, Karlsruhe Institute of Technology, Karlsruhe, Germany

Jan Seidel
School of Materials Science and Engineering, University of New South Wales, Sydney, Australia

Qisheng Yu
Key Laboratory for Quantum Materials of Zhejiang Province, Department of Physics, School of Science, Westlake University, Hangzhou, Zhejiang, China

List of abbreviations

AFM	Atomic force microscopy
BPNN	Behler–Parrinello neural network
BPVE	Bulk photovoltaic effect
BVS	Bond-valence sum
BVVS	Bond-valence vector sum
CBM	Conduction band minimum
CMOS	Complementary metal-oxide-semiconductor
DC	Direct current
DFT	Densify functional theory
DNN	Deep neural network
DP	Deep potential
DP-GEN	Deep potential generator
DPMD	Deep potential molecular dynamics
DW	Domain wall
EBSD	Electron backscatter diffraction
FA	Formamidinium
FePV	Ferroelectric or polarization-assisted photovoltaic effect
GDML	Gradient-domain machine learning
IMH	Inorganic metal halides
IoT	Internet of Things
KPFM	Kelvin probe force microscopy
LGD	Landau–Ginzburg–Devonshire
MA	Methylammonium
MD	Molecular dynamics
ML	Machine learning
MPB	Morphotropic phase boundary
NN	Neural network
OMH	Organic metal halide
PCA	Principal components analysis
PCE	Power conversion efficiency
PFM	Piezoresponse force microscopy
PhFM	Phenomenological phase-field model
P-LED	Perovskite light emitting device
PZT	$Pb(Zr,Ti)O_3$
REBO	Reactive empirical bond-order
SEM	Scanning electron microscopy
SHG	Second-harmonic generation
SOJT	Second-order Jahn–Teller
SRH	Shockley–Read–Hall
STEM	Scanning transmission electron microscopy
SVD	Singular value decomposition
VBM	Valence band maximum
vdW	van der Waals
WSN	Wireless sensor networks
XCD	Additional circular dichroism
XRD	X-ray diffraction

Chapter 1

Introduction

Yang Bai and Ilya Grinberg

This opening chapter overviews the structure and content of this book. Essential terms regarding ferroelectrics are briefly introduced. The rationale for the selected topics in this book in line with current focus of research activities is summarized and relationships among different chapters are explained. This chapter will help readers to quickly navigate suitable contents to proceed further based on the readers' interests and backgrounds.

It has been said that ferroelectricity cannot be predicted from the crystal structure of the materials and must be validated by experiments (Moulson and Herbert 2003). With the development of modelling and simulation methods in the recent decade, predictive ability is being established quickly. For this reason, part I of this book discusses the advances in the fundamental understanding of ferroelectricity that can be used to reliably predict and lead and assist design of new ferroelectric materials. A tremendous number of new ferroelectrics are now urgently needed. This is because ferroelectricity induces multiple degrees of freedom that enable the operation of devices made from these materials. Each specific application may require a different combination of, e.g. strength of polarization, coercive field, switchability, crystal structure and boundary, Curie temperature and phase transition, functional properties, etc. The needed rapid device innovations cannot afford the conventional, tedious experimental procedures that have been used for nearly a century to discover and study new ferroelectric compounds. This firmly links part I to part II, which discusses emerging applications of ferroelectric materials and devices enabled by theoretical prediction of new ferroelectrics.

1.1 Spontaneous polarization and phase transition

Ferroelectric materials are often analogized to ferromagnetic counterparts to ease the understanding of ferroelectricity. However, the only analogy here is the switching behaviour, i.e. in ferromagnets the magnetization can be switched by external magnetic field while in ferroelectrics the polarization can be switched by external

doi:10.1088/978-0-7503-3975-9ch1

electric field. Spontaneous polarization, which is induced and exists spontaneously in the unit cells of a material, is the first criterion to define ferroelectricity. Figure 1.1 gives an example of spontaneous polarization in an oxide perovskite unit cell (ABO_3) where A-site cations occupy the corners of the cube, the B-site cation sits on the body centre and the oxygen ions take the face centres. Below the Curie temperature (T_C), the BO_6 octahedron naturally distorts by shifting the B-site cation away from high-symmetry position. This results in the separation of the cation and anion centres and gives rise to a spontaneous polarization.

The example shown in figure 1.1 is for the cubic (C)–tetragonal (T) phase transition, where in the C phase spontaneous polarization is forbidden and in the T phase the spontaneous polarization is along the 001 direction, i.e. pointing from the horizontal plane of the BO_6 octahedron vertically upwards and the B-site cation shifts along this direction. Generally, in the T phase six directions are allowed to host the spontaneous polarization, namely the 001, 010, 100 and their corresponding opposite directions. Depending on the composition of the oxide perovskite, e.g. $BaTiO_3$, $PbTiO_3$, etc as listed in figure 1.1, different phase transitions may occur across and further below T_C. The features of the oxide perovskite unit cells and their phase transitions and the resultant ferroelectricity have been well observed experimentally for different compositions. The empirical evidence has provided a database for the development of theories. Therefore, chapter 2 of this book discusses the development of first-principles calculation and density functional theory studies used to predict the unit-cell level properties such as spontaneous polarizations, as well as phase transitions and collective properties for complex ferroelectric solid solutions based on the oxide perovskite structure. Chapter 3 extends the prediction to multiscale methods, including the state-of-the-art machine learning methods. Collaborative use of first-principles calculations and multiscale simulations can reliably predict a comprehensive set of ferroelectric properties. Chapter 4 provides an example of predicting ferroelectric behaviours under different external boundary conditions caused by domains and domain walls.

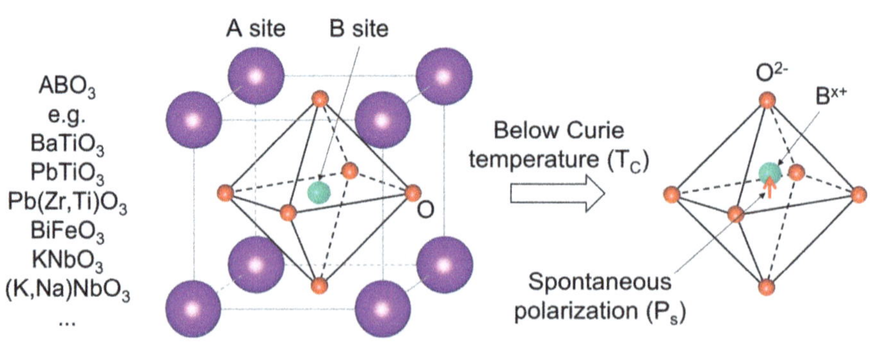

Figure 1.1. Schematics of the oxide perovskite unit cell (ABO_3) and spontaneous polarization. Note that the A-site cations are not shown in the schematic on the right for a better visual presentation.

1.2 Domain, domain wall and ferroelectric switching

Although there are multiple allowed directions in a given phase for spontaneous polarizations to form, switching the already established individual spontaneous polarization in a unit cell to another direction is only theoretically possible. In practice, ferroelectric polarization is switched through domains and mostly via domain wall motion. Figure 1.2 schematically explains that a group of unit cells that possess spontaneous polarization in the same direction forms a domain. In the same domain, the polarization direction is the same but the adjacent domains in a crystal may have polarization oriented in different directions. The interface between two domains having different polarization directions is called a domain wall. If an external electric field is applied, some domains will reorient and align with the direction of the electric field.

Figure 1.3 illustrates the evolution of domains and domain walls when subject to an external electric field (*E*), for example, in a T ferroelectric phase where 180° and 90° domain walls are present. When the electric field is applied, the domains with

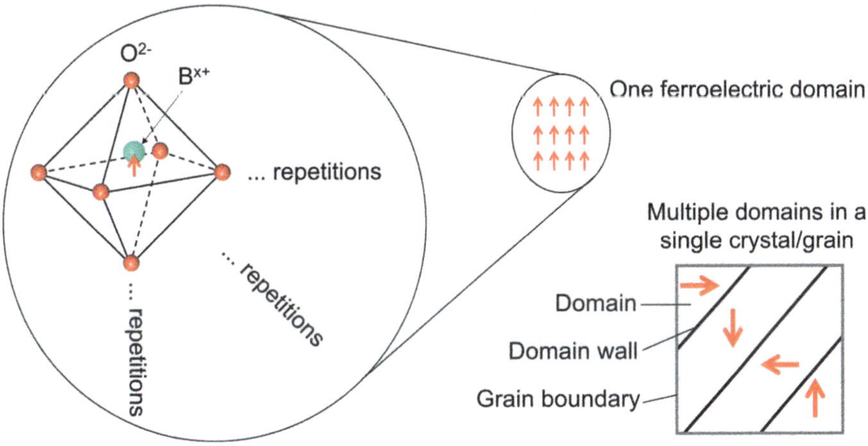

Figure 1.2. Schematics of domains and domain walls. A-site cations are hidden for simplicity.

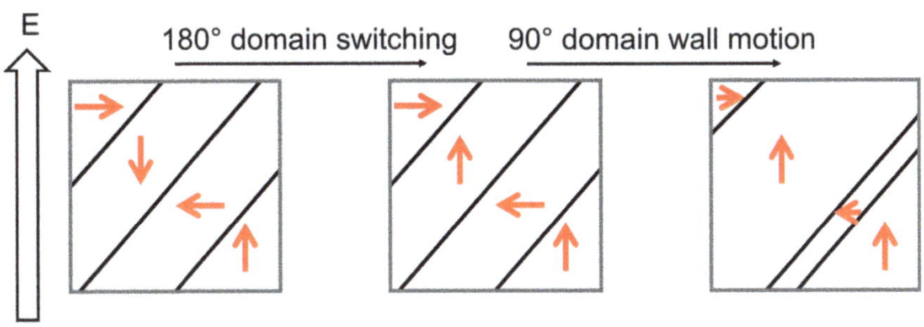

Figure 1.3. Schematics of domain switching and domain wall motion. *E* represents external electric field.

polarizations in parallel with but opposite to the electric field may switch by 180° to be aligned with the field. On the other hand, the already aligned domains may expand under the electric field by forcing a part of the spontaneous polarizations in the adjacent 90° domains to switch. Thus, new domain walls are formed between the switched and unswitched areas in the pristine 90° domains. This creates the appearance that the 180° domains are consuming the 90° domains so that the term domain wall motion is assigned to this process.

The extent of domain switching and domain wall motion depends on the strength of the external electric field. Figure 1.4 shows a standard ferroelectric hysteresis loop (also called a *P–E* loop) indicating the relationship between the net polarization of a ferroelectric material and external electric field applied to the material. A pristine, unpoled ferroelectric does not exhibit any net polarization (point 0 in figure 1.4) since the domains are randomly oriented. Driven by electric field, a part of the domains starts to reorient and the larger the electric field, the more domains are reoriented (0→1 in figure 1.4) until saturation is reached (point 1). In this state, majority of the domains have been permanently switched and thus a remanent polarization (point 2) is left after removing the electric field (1→2). The remanent polarization can be erased by applying a negative electric field until the coercive field (2→3) and the net polarization returns to zero (point 3). Further increasing the negative electric field (3→4) will bring the net polarization to the saturation in the other direction (point 4) and from there the procedure (4→5→6) is repeated through the other remanent polarization (point 5) and finishes at the positive coercive field with yet another zero net polarization reached (point 6). The loop (1→6) will replicate itself after the first cycle and the shape of the loop is affected by frequency of the electric field as well as material properties. Note that the value of the saturated polarization is not necessarily the same as that of the spontaneous polarization.

Research on ferroelectrics has been strongly focused on the manipulation of domains, domain walls and their switchability. Not only can the spontaneous polarizations and coercive fields be predicted (chapters 2 and 3), but also the

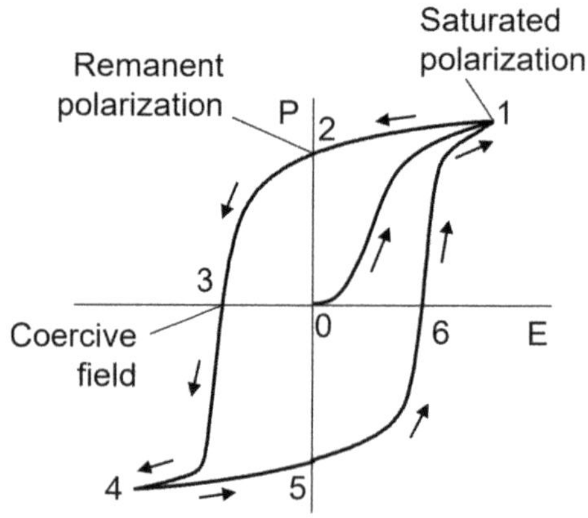

Figure 1.4. Example ferroelectric hysteresis loop (*P–E* loop). *P* is polarization.

switching behaviour can be estimated by calculation as will be discussed in chapter 4. Apart from electric field, domain wall motion may also be triggered by stress and temperature. Chapters 2–4 will explain such behaviours. Domain walls are in fact topological defects that relax the strain or other interfacial energy between two domains. Although called defects, they do not necessarily degrade functionality but rather offer fascinating opportunities for fundamental and experimental studies. Chapter 6 connects the theories of domain walls to experiments and discusses the features of topological defects in ferroelectrics.

Conventional ferroelectric research has not been focused on light as a potential driving force for changes of the ferroelectric material. However, research in the recent decade has increasingly shown that light may also play a role similar to that of electric field, stress or temperature for operation of ferroelectrics. Therefore, chapters 4 and 5 collaboratively discuss the interaction between light and ferroelectricity from the theoretical and experimental points of view, respectively. In particular, chapter 5 discusses the multi-dimensional behavior involving opto-ferroelectric effects as well as the coupling of photon energy with piezoelectric (electric field–stress coupling, namely electromechanical coupling) and pyroelectric (electric field–temperature coupling, namely electro-thermal coupling) effects.

1.3 Piezoelectric and pyroelectric effects

Materials at the status of point 2 or 5 shown in figure 1.4 are called poled ferroelectrics. Poled ferroelectrics exhibit piezoelectric and/or pyroelectric effects. The piezoelectric effect is defined as the generation of electric charge displacement upon material deformation, e.g. caused by stress or strain (direct piezoelectric effect) and vice versa, the generation of material deformation upon application of electric field (converse piezoelectric effect). Pyroelectric effect represents the phenomenon of electric charge displacement generated upon the change of material temperature. Both piezoelectric and pyroelectric effects are also eventually a consequence of domain wall motion but without the material being fully depoled (i.e. working in the $1 \rightarrow 2$ or $4 \rightarrow 5$ region). Piezoelectric and pyroelectric devices are widely used as sensors, actuators and transducers. Their applications in energy harvesting have grown rapidly in the recent decade. For this reason, chapter 5 also introduces the development of energy harvesting by emphasizing the multi-functional features of ferroelectrics.

1.4 Halide perovskites and ferroelectric microelectronics

The ABO_3 structure shown in figure 1.1 is closely related to the structure of halide perovskites in which O^{2-} is replaced by Cl^-, Br^-, I^-, etc. This difference, together with other changes on the A and B sites, greatly broadens the applications of the perovskite family. More importantly, in the recent decade halide perovskites have been shown in some works to also be ferroelectric. Experience obtained with oxide perovskites is being transferred to halide perovskites and the new discoveries made with halide perovskites are feeding back to the development of oxide perovskites. The mutual use of oxide and halide perovskites will help to explore new types of microelectronics in the fields of sensors, smart devices and energy technology. Therefore, chapter 7 ends this book by discussing ferroelectricity in halide perovskites and chapters 5–7 combine the discussion of emerging applications of ferroelectrics in microelectronics.

Reference

Moulson A J and Herbert J M 2003 *Electroceramics: Materials, Properties, Applications* 2nd edn (Chichester: Wiley) p 18

Part I

Fundamental evolutions of ferroelectricity

IOP Publishing

Ferroelectrics
Advances in fundamental studies and emerging applications
Yang Bai and Ilya Grinberg

Chapter 2

Theoretical advances in understanding complex ferroelectrics

Ilya Grinberg

This chapter describes the theoretical understanding of complex ferroelectric solid solutions achieved over the past several decades. Ferroelectrics with the best performance are usually multicomponent solid solutions for which compositional variation is used as a tool to adjust the interplay of various interatomic interactions in order to obtain the best response to stimuli such as stress and electric field. In the past three decades, first-principles methods have been used in an ongoing effort to elucidate how the composition and interatomic interactions in complex ferroelectric perovskite oxides give rise to their excellent properties. First starting with a simple five-atom system and focusing on the basic properties such as atomic positions and polarization, these studies have advanced to investigations of solid solutions with multiple elements on the perovskite A- and B-sites and collective properties such as transition temperature and coercive field and obtained accurate composition–structure–property relationships to guide further experimental materials design studies.

2.1 Brief history and overview of complex ferroelectrics

Ferroelectrics are materials with a spontaneous switchable polarization that play a key role in a variety of device applications such as in dielectrics and piezoelectrics. In addition to their technological importance, they are also fundamentally fascinating materials because the presence of spontaneous polarization allows a wide variety of local structures that are sensitive to the changes in the composition, structure and environment of the material. Both academic and industrial interest in ferroelectrics has been focused mostly on perovskite (ABO_3) oxides, starting with the prototypical classic $BaTiO_3$ material first discovered in the 1940s. Over the last eight decades, many new perovskite ferroelectric (FE) materials have been synthesized and explored. The discovery of $BaTiO_3$ in the 1940s (von Hippel 1950) with its complex phase transition sequence of rhombohedral (R)–orthorhombic (O)–tetragonal (T)–cubic (C) with increasing temperature was followed by

doi:10.1088/978-0-7503-3975-9ch2

the synthesis and discovery of $PbTiO_3$ in 1950 (Shirane *et al* 1950) that shows only a simple T–C phase transition as well as a high transition temperature (T_C) of 765 K. In the exploration of solid solutions of $PbTiO_3$, it was quickly realized that the highest dielectric and piezoelectric properties can be obtained by mixing $PbTiO_3$ with other $Pb(Me)O_3$ materials (Me represents the metal cations occupying the perovskite B-site, e.g. Me=Zr, $Mg_{1/3}Nb_{2/3}$, $Mg_{1/2}W_{1/2}$, $Sc_{2/3}W_{1/3}$, etc). In particular, by mixing $PbTiO_3$ with $PbZrO_3$ at 52/48 composition, the famous PZT ($PbZr_{1-x}Ti_xO_3$) material with high piezoelectric performance was obtained (Jaffe *et al* 1954) that 70 years later still forms the basis of mass-produced piezoelectrics (figure 2.1). This launched the still-ongoing effort to understand and exploit these multicomponent ferroelectric oxide materials.

Multicomponent ferroelectric perovskite oxides are solid solutions and are intrinsically complex materials. A simple material such as $PbTiO_3$ can be easily characterized by specifying the coordinates of the atoms in its five-atom unit cell, shows long-range order and usually is fairly insensitive to applied external perturbations such as stress or temperature. By contrast, microscopically the Pb $(Zr_{1-x}Ti_x)O_3$ solid solutions display a random arrangement of Zr and Ti atoms on the perovskite B-site, as well as Å-scale spatial variations in the magnitude and orientation of cation displacements from the high symmetry structure. Macroscopically, the PZT phase diagram exhibits a rich variety of phases ranging from the AFE (antiferroelectric) phase on the Zr-rich side, to the R, M (monoclinic) and then T phases as Ti content is increased (figure 2.1(a)). The high-temperature phase is always the paraelectric C phase, but the phase transition temperatures show non-monotonic variation with the Zr/Ti content. Additionally, the dielectric and piezoelectric response of the structure shows a spike at the morphotropic phase boundary (MPB) occupied by the M phase (figure 2.1(b)) (Jaffe *et al* 1954, Du *et al* 1998). As

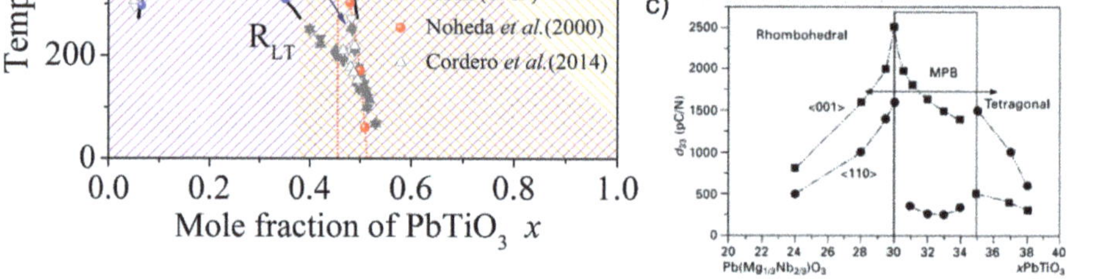

Figure 2.1. (a) Phase diagram of PZT according to experimental measurements. Reprinted from Lu *et al* (2016). Copyright © 2016 AIP Publishing. (b) Dependence of dielectric and piezoelectric constants on PZT composition, showing a spike at 50/50 composition at the MPB. Reprinted from Du *et al* (1998). Copyright © 1998 AIP Publishing. (c) Dependence of piezoelectric response on composition in the PMN-PT, showing a spike at the 33/67 composition at the MPB. Reprinted from Luo *et al* (2010). Copyright © 2010 Elsevier.

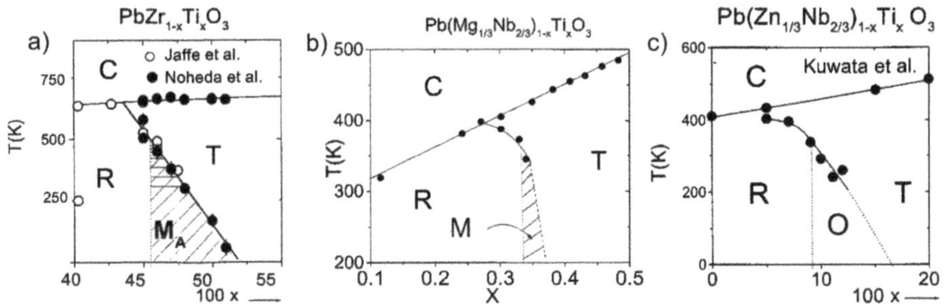

Figure 2.2. Phase diagrams of PZT, PMN-PT and PZN-PT showing the universal sequence of phase transitions for $PbTiO_3$ solid solution. (a), (c) Reprinted from Cox *et al* (2001). Copyright © 2000 AIP Publishing. (b) Reprinted from Ye *et al* (2001). Copyright © 2001 American Physical Society.

shown in figures 2.1(c) and 2.2, similar phase diagrams are also found for other solid solutions with $PbTiO_3$ such as the famous $(1-x)Pb(Mg_{1/3}Nb_{2/3})O_{3-x}PbTiO_3$ (PMN-PT) system with similar spikes in the response properties (Noheda *et al* 1999, 2001, Guo *et al* 2000, Ye *et al* 2001, Cox *et al* 2001, Luo *et al* 2010). Moving from binary to ternary and quaternary solid solutions e.g. $(1-x-y)Pb(Mg_{1/3}Nb_{2/3})O_{3-y}$ $Pb(In_{1/2}Nb_{1/2})O_{3-x}PbTiO_3$ (PIN-PMN-PT) increases the complexity of the structure but allows one to obtain improved properties (Tian *et al* 2007, Zhang *et al* 2008).

Relaxor behavior is another well-known example of the complexity found in ferroelectric perovskites. Relaxor ferroelectrics are solid solutions with multiple B-site and/or A-site elements that display diffuse phase transitions as well as dispersion of the dielectric response with frequency. While they show sharp x-ray diffraction (XRD) peaks due to the preservation of the overall perovskite structure, since the seminal work of Burns and Dacol in 1983 relaxors have been known to exhibit inhomogeneities on multiple length scales related to the orientations of local dipoles formed by cation off-center displacements and possibly local (Å- or nm-scale) compositional variations (Burns and Dacol 1983, Fu *et al* 2009, Tsukada *et al* 2017, Takenaka *et al* 2017, Eremenko *et al* 2019).

It is logical to suppose that it is precisely the complexity and the Å-scale inhomogeneity of the composition and structure of these solid solutions that gives rise to their rich macroscopic properties; however, complexity leads to the challenge in relating the material composition to its properties. This challenge has motivated research over the last several decades that has made significant progress in understanding these complex materials.

In the quest to obtain theoretical understanding of complex ferroelectrics, it is necessary to seek to elucidate simple rules that can accurately characterize and predict how the composition of these materials is related to their structure and then to their properties, both micro- and macroscopic, based on the fundamental physical and chemical principles and the interatomic interactions in these materials. While both experimental methods and first-principles calculations can provide the information about a material's composition, structure and properties, neither the experimental results nor the energies and atomic positions obtained by first-principles calculations in and of themselves provide information about the composition–structure–property relationships that then can be used to explain the properties of a wide range of materials and design new materials with the desired target properties. Therefore, for a theoretical understanding of these complex systems, the generality of first-principles calculations must be sacrificed and composition–structure–property relations based on the application of the general knowledge about chemical bonding and physical interactions to the specific case of ferroelectric perovskite solid solutions must be developed. For this purpose, the key properties of the solid solutions such as the characteristics of the constituent atoms, and their relationship to the key interatomic interactions must be identified. To summarize the complex compositional and structural variations in solid solutions, it is natural to use the approach of using average properties and the extent of their variation as the descriptors of the composition and structure of the solid solution. As will be shown below, such elucidation of the descriptors then allows us to describe and predict key properties of ferroelectric oxides.

2.2 Fundamental chemistry and physics of ABO_3 ferroelectrics

Ferroelectricity in perovskite oxides arises from the spontaneous displacements by the A-site and B-site cations from the center of their oxygen complexes (O_6 for B-site, O_{12} for A-site). Second-order Jahn–Teller (SOJT) distortion is the mechanism that gives rise to the preference for a distorted structure of the B–O_6 octahedron (Wheeler *et al* 1986). The SOJT distortion arises due to the fact that the overlap between the d orbitals of the B-site cation and the 2p orbitals of the O atoms is overall non-bonding in the undistorted structure (figure 2.3(a)), whereas an off-center displacement by the B-cation brings the B-cation closer to the O atom on one side and further from the O atom on the other side. This makes the constructive 3d–2p orbital overlap greater than the destructive 3d–2p orbital overlap and lowers the energy of the structure. This means that the presence of d orbitals and their hybridization with the 2p orbitals is necessary for ferroelectricity to occur, as was described by Cohen (1992) in his seminal paper on the 'Origin of ferroelectricity in

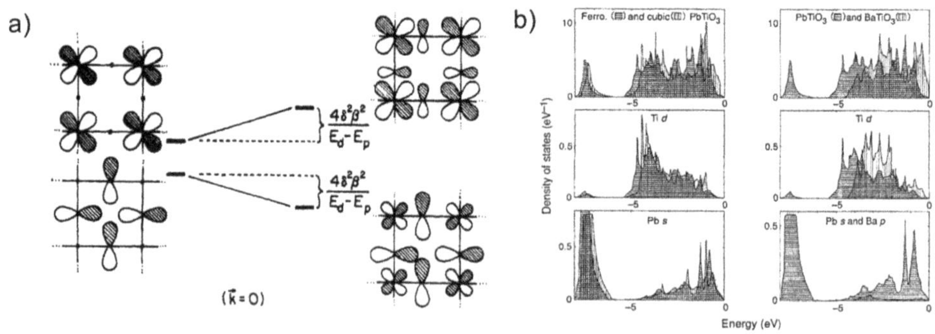

Figure 2.3. (a) Second-order Jahn–Teller distortion mechanism that gives rise to off-center B-site displacement in ferroelectric perovskite oxides. Reprinted from Wheeler *et al* (1986). Copyright © 1986 American Chemical Society. (b) Densities of states from DFT calculations showing hybridization between TM d and O2p orbital that gives rise to the SOJT effect. Reprinted from Cohen (1992). Copyright © 1992 Springer Nature.

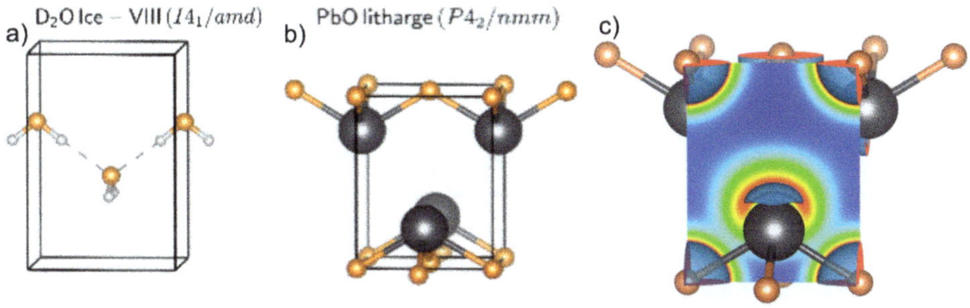

Figure 2.4. (a) Structure of ice showing the distorted structure of water molecule due to the presence of lone pairs on the O atoms. (b) Structure of PbO showing local distortions similar to those of water due to the stereochemically active lone pair on the Pb atom. (c) Charge density showing the lone pair on the Pb atom. Reprinted from Laurita *et al* (2022). Copyright © 2022 American Chemical Society.

perovskite oxides' (figure 2.3(b)). For the A-site, the energetic preference for the off-center position is driven by the *sp* lone pair that is present in the Pb and Bi elements and that creates a fundamentally asymmetric local environment for the Pb and Bi elements (figure 2.4) (Laurita and Seshadri 2022). This is similar to the asymmetric structure of H_2O that is created due to the presence of lone pairs on the O atom (figure 2.4). Therefore, large off-center A-site displacements are observed in perovskite oxides with Pb and Bi on the A-site, while much smaller displacements are observed for the simple metal cations Ba, K, Sr, Na that are spherically symmetric and therefore prefer the high-symmetry A-site position.

Another important factor that must be taken into account when considering ABO_3 materials is the degree of matching between the preferred A–O and B–O bond lengths as characterized by the well-known Goldschmidt tolerance factor (Goldschmidt 1926). Due to the cubic geometry of the ABO_3 cells, the ratio between the A–O bond length (one half of the cube face diagonal) and the B–O bond length (one half of the cube edge) is equal to $\sqrt{2}$. Therefore, no bond strain will be present in the system when the tolerance factor t defined as

$$t = (R_A + R_O)/(R_B + R_O)\sqrt{2} \tag{2.1}$$

is equal to 1, where R_A, R_B and R_O are the ionic sizes of the A, B, and O atoms. If t is less (greater) than 1, this means that the A-site prefers a smaller (larger) bond length than the B-site. To minimize the bond strain, the ABO_3 structure will distort either by tilting the O_6 octahedra for $t < 1$ or by displacing the B-site ion for $t > 1$. Thus, the ground state structure of a perovskite is based on a compromise between the preferences of the A–O and B–O sublattices.

In addition to the A–O and B–O interactions that determine the preference of the A- and B-site cations for off-center displacement, the A–A, B–B and A–B interactions also play an important role in determining the energy of a given ABO_3 structure. Since there are no A–B covalent bonds in perovskite oxides, the interactions between A- and B-cations will be repulsive. This short-range repulsion can be quite strong and scales with the orbital overlap of the A- and B-cation

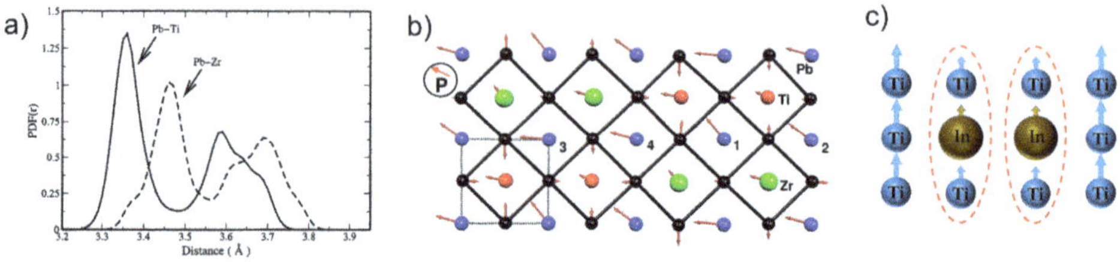

Figure 2.5. (a) Pb–Ti and Pb–Zr pair distribution functions showing the larger Pb–Zr distance compared to Pb–Ti distance. Reprinted from Grinberg *et al* (2004). Copyright © 2004 American Physical Society. (b) Pb atoms avoid Zr and displace toward Ti. Reprinted from Grinberg *et al* (2002). Copyright © 2002 Springer Nature. (c) Coupling between the cation displacements due to the bond-valence conservation effect. Reprinted from Yadav *et al* (2022). Copyright © 2022 Wiley.

orbitals (Grinberg *et al* 2002, 2004). This means that larger A–B site distances will be preferred for larger A–B site atoms. Thus, the Pb–Zr distance in PZT will be larger than the Pb–Ti distance (figure 2.5(a)) as the Pb atoms will avoid Zr atoms and displace toward Ti atoms (figure 2.5(b)).

Since the nearest A(B) site neighbors of a A(B) site are separated by O atoms, there are no A–A and B–B short-range repulsions. However, the displacements of the A- and B-sites in the neighboring unit cells are coupled by through-oxygen interactions due to bond-order conservation. Starting in the 1970s, analysis of crystal structures showed that in a stable structure, all atoms are close to their ideal valence, e.g. 4 for Ti and 2 for Ba and O in $BaTiO_3$ (Brown 2009). The total valence of an atom i (V_i) is the sum of the individual bond valences given by

$$V_i = \sum_j^N \exp\left(-(R_{0,ij} - R_{ij})/b\right) \tag{2.2}$$

where $R_{0,ij}$ is a constant specific to each pair of elements of atoms i and j, b is a universal constant typically taken as 0.37 and R_{ij} is the distance between the atoms i and j. It can be seen that the relationship between bond length and bond order is exponential as was first noted by Pauling due to the key role of orbital overlap in bonding. Therefore, an atom with a higher (lower) valence is overbonded (underbonded) and this will lead to a higher energy of the structure. This principle was later used for parameterization of atomistic force fields for ferroelectric perovskite as described in more detail in chapter 3. This means that a displacement of a cation toward an O atom necessitates the displacements of other cations away from this O atom due to the need to compensate the increased bond order of the shorter bond length by the decrease in the bond order of the longer bond (figure 2.5(c)). This favors the coordinated displacement of the A- and B-site cations.

The basic interactions described above: second-order Jahn–Teller distortion and lone-pair effects that give rise to the preference of off-centered structures, the bond strain due to the mismatch of the A–O and B–O bond lengths, the through-oxygen coupling due to bond order conservation and the short-range A–B repulsion, and their interplay explain the relationship between the composition, structure and properties of complex ABO_3 solid solutions, as elucidated by theoretical and experimental investigations over the past two decades, as described below.

2.3 Understanding the basic ordered five-atom structure

First, we discuss the ordered ABO_3 ferroelectrics to illustrate how the interactions described above can be used to explain the compositional trends in these systems. In the framework of perturbation theory, the energy lowering ΔE_{FE} due to the more favorable mixing of the TM (transition metal) d and O 2p orbitals in the off-center distorted structure should scale according to

$$\Delta E_{FE} = V_{TM-O}^2/(\varepsilon_O - \varepsilon_{TM}) \quad (2.3)$$

where ε_O and ε_{TM} are the energies of the TM d and O2p orbitals and $V_{TM\text{-}O}$ is the hopping matrix element (Wheeler *et al* 1986). This simple argument can be further improved by taking into account the effect of the short-range repulsion between the TM cation and oxygen in the framework of vibronic-coupling theory (Polinger *et al* 2015, Bersuker 2018). Taking the short-range repulsion and SOJT effects into account, it was shown (Yadav *et al* 2022) that the square of the expected ionic displacement is proportional to the vibronic-coupling matrix element V_{TM-O} and the crystal stiffness that can be reflected by ionic radius of the B-cations $R_{B,}$ and to the transition metal–oxygen states' eigenvalue difference reflected by the electronegativity of the cation as expressed in

$$D_0^2 = C_0 + C_1 R_B^{-1} + C_2 \chi_B \quad (2.4)$$

where R_B and χ_B are the ionic radius and electronegativity of the B-site transition metal ion and c_0, c_1 and c_2 are constants. This equation explains the empirically observed trends. For example, $KNbO_3$ is a strongly ferroelectric material with a T_C of 705 K. Substitution of Nb by Ta that has the same valence and ionic size as Nb obtains the $KTaO_3$ quantum paraelectric material with a $T_C \approx 0$ K. This rather large effect is due to the difference in the $(\varepsilon_O - \varepsilon_{TM})$ between Nb and Ta. Such difference can be measured by the band gap of the binary oxide (Nb_2O_5 versus Ta_2O_5) because across-the-gap excitation in oxides is equivalent to charge transfer from the O2p states to the TM d states (Eng *et al* 2003). Thus, the band gap measures the eigenvalue differences. The band gap of Ta_2O_5 is ~1 eV higher than that of Nb_2O_5, dramatically decreasing the driving force for distortion and making $KTaO_3$ non-polar. Similarly, ions with larger size tend to show smaller displacements in ABO_3; this is directly related to their larger size and greater crystal stiffness. As shown in figure 2.6, this relationship predicts well the displacement magnitude found by density functional theory (DFT) calculations for various B-site TM cations in ABO_3 solid solutions, and it also predicts the displacement magnitudes for the sp-block B-site cations.

Proceeding from the unit-cell level properties of individual cations, the arrangement of the cations is the next feature of the material that is important for its properties. The factors controlling the ordering of cations in ABO_3 solid solutions were investigated thoroughly in the late 1990s, most importantly by Davies and

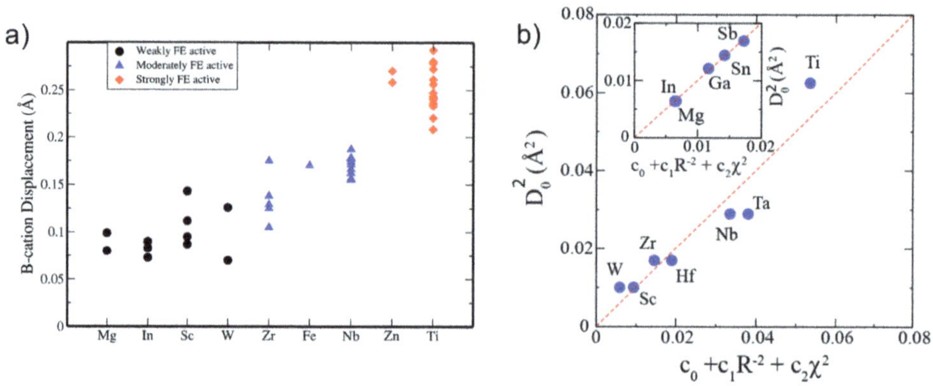

Figure 2.6. (a) Displacements of different B-site cations in $PbTiO_3$-rich solid solutions obtained from DFT calculations. Reprinted from Grinberg *et al* (2005). Copyright © 2005 AIP Publishing. (b) Average displacement values from (a) predicted based on the ionic size and electronegativity of the cations. Reprinted from Yadav *et al* (2022). Copyright © 2022 Wiley.

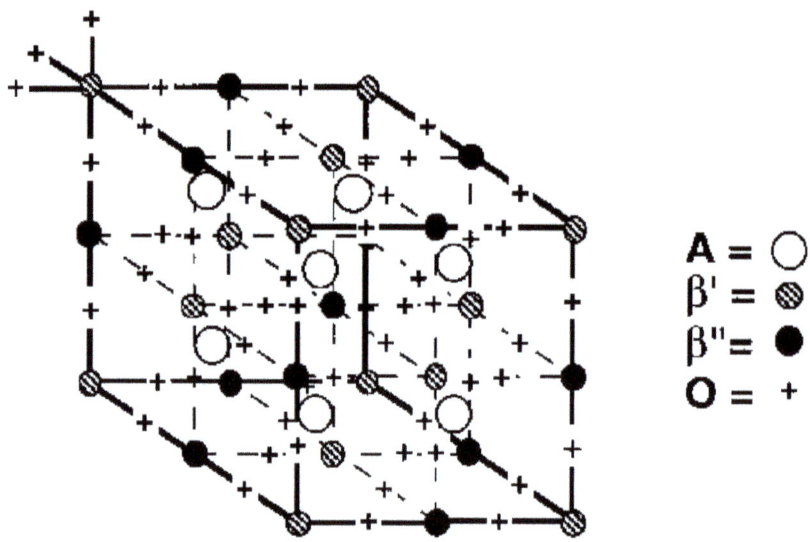

Figure 2.7. Schematic illustration of the rock-salt arrangement of the β′and β″ B-sites. Reprinted from Davies and Akbas (2000). Copyright © 2000 Elsevier.

co-workers experimentally (Akbas and Davies 1997, Davies 1999, Farber and Davies 2003) and by Burton and Cockayne theoretically (Burton and Cockayne 1999). ABO_3 solid solution with B-site cations that are different in charge and size can display long-range rock-salt (RS) ordering. An example of such ordering is shown for the $BaSc_{0.5}Nb_{0.5}O_3$ material in figure 2.7. A charge difference of 4 (e.g. in $PbMg_{0.5}W_{0.5}O_3$ known as PMW) always leads to a spontaneously ordered structure with the RS arrangement of +2 and +6 cations on the B-site. This is due to the larger electrostatic driving force for order. (Here, the +2 and +6 cations can be thought of as −2 and +2, respectively, relative to the average B-site charge. It is obvious that this leads to the preference for the NaCl structure.) A charge difference of zero (e.g. for PZT or $KNb_{1-x}Ta_xO_3$ known as KTN) will always lead to a disordered arrangement of the cation. For a charge difference of 2 (e.g. $BaSc_{0.5}Nb_{0.5}O_3$ and $PbSc_{0.5}Nb_{0.5}O_3$ known as BSN and PSN, respectively), some solid solutions display long-range-ordered $\beta'\beta''$ arrangements (figure 2.7) while others show no long-range order. Davies and co-workers showed that the ordering is favored by the greater difference between the β' and β'' cation sizes because in this case the RS arrangement also allows each β'-O6 and β''-O6 octahedron to achieve its preferred bond lengths because the large and small cations will be always matched, whereas this is not the case in a disordered structure (Davies 1999).

It was also shown (Davies 1999) that $Pb(B'^{+2})_{1/3}(B''^{+5})_{2/3}O_3$ and $Pb(B'^{+3})_{2/3}(B''^{+6})_{1/3}O_3$ perovskites adopted the random-site ordered structure (figure 2.8) where one site was fully occupied by the majority B-site cation while the other site was occupied by a mixture of the majority and the minority B-site cations to achieve an average B-site difference of +2. Such random-site RS order was favored by the larger difference in the cation ionic size between the majority site and the random site (using the average cation radius of the random site). For example, $PbFe_{2/3}W_{1/3}O_3$ (PFW) does not show long-range B-cation site order while $PbSc_{2/3}W_{1/3}O_3$ (PSW) does. In an interesting example of the effects of such B-cation ordering, it was found that the substitution of the strongly ferroelectric $PbTiO_3$ (T_c = 765 K) into the relaxor PSW ($T_c \approx 250$ K) counterintuitively leads to a decrease in T_c (Juhas *et al* 2004a, 2004b).

In a key theoretical study, Burton and Cockayne (1999) noted that while Pb-based perovskites order with difficulty, B-cation ordering is much easier to obtain in their Ba-based analogs, as can be seen from the comparison of the long-range RS-ordered BSN to PSN that shows only weak long-range RS ordering without the use of slow cooling during synthesis. In first-principles calculations, this is manifested by

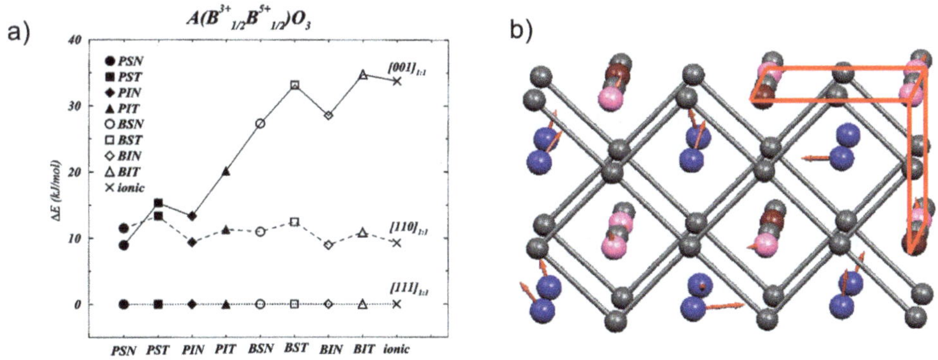

Figure 2.8. (a) Energy differences between the RS and other B-cation arrangements for different perovskite oxides. Reprinted from Burton and Cockayne (1999). Copyright © 1998 American Physical Society. (b) Random-site ordered structure of 60-atom PMN supercell obtained from DFT calculations. Pb atoms displace toward underbonded Mg–O–Nb atoms and away from overbonded Nb–O–Nb atoms. Reprinted from Grinberg and Rappe (2004). Copyright © 2004 American Physical Society.

the smaller energy differences between the RS-ordered B-cation arrangement and other cation arrangements for PSN compared to BSN (figure 2.8(a)). This was explained using the concept of under- and overbonding. As mentioned above, each O atom seeks to make two bonds. In a Ba or Pb-based ABO_3, the average valence of the B-site must be 4. Therefore, an O ion with two Sc (Nb) neighbors has less (more) than required number of bonds with the B-site cations and is underbonded (overbonded). This deviation from the ideal bonding configuration can be compensated by the distortions of the A-site cations. An underbonded O atom can compensate by creating stronger bonds with the A-site cations through the displacement of the A-site cation toward the underbonded oxygen. Conversely, an overbonded O atom can compensate by creating weaker bonds with the A-site cations through the displacement of the A-site cation away from the overbonded oxygen (figure 2.8(b)). Such compensation is easier to achieve in Pb-based perovskite due to the large intrinsic displacement of the Pb arising from the presence of a stereochemically active lone pair as described above. By contrast, the preference of Ba to remain in the center of its $A\text{–}O_{12}$ complex imposes an energetic cost for a distortion that alleviates the over- or underbonding of the O atoms. Therefore, the energy cost of disordering in Ba-based perovskites is larger and they order more easily (for lower ionic size difference between the β' and β'' cations $\Delta R_{\beta'-\beta''}$ and without slow cooling) while for the Pb-based perovskites larger $\Delta R_{\beta'-\beta''}$ differences and slow cooling are necessary, as can be seen by the comparison of BSN and PSN. Thus, Burton and Cockayne made an important contribution by pointing out that the driving force for ordering is not purely electrostatic but has a strong covalent component as well, based on the bond valence conservation principle. It therefore follows that Bi-based perovskites will show even weaker driving force for ordering due to the very high displacement of Bi that will allow Bi displacements to better compensate the O under- and overbonding created by the disordered B-cation arrangement. This was in fact found experimentally where the bulk $Bi(Mg_{1/2}Ti_{1/2})O_3$ (BMT) synthesized at high pressure showed no evidence of long-range cation order (Khalyavin *et al* 2006) in contrast to PSN which has a smaller $\Delta R_{\beta'-\beta''}$ (0.10 Å) than that of BMT (0.12 Å) and the same charge difference and can show long-range cation order as discussed above. Thus, in ferroelectric solid solutions, the A-site can modulate the ordering tendencies of the B-site through bond order compensation to achieve bond valence conservation (figure 2.8(b)).

2.4 Compositional phase transitions

Phase information and lattice parameters are perhaps the most basic characteristics of a crystal. For ferroelectric solid solutions, phases also show a one-to-one relationship to the orientation of the polarization (P) vector (due to strain–polarization coupling). Thus, the P vector is oriented along the (111), (110) and (100) cubic directions for the R, O and T phases. For the M phase, the P vector can be along the plane from (111) to (100) (M_a), or the plane from (110) to (100) (M_c). The cubic phase is non-polar with no static polarization. Therefore, phase diagrams provide the most basic information about solid solutions and the phase transitions are correlated with the high response properties of FE materials; thus, understanding of how composition controls the phase diagram is the first step in understanding the properties of these complex solid solutions. The phase diagram of PZT is a classic example. PZT shows a sequence of compositional phase transitions from AFE to R FE, M FE, and T FE (figure 2.1). In its analogs created by the replacement of $PbZrO_3$ end-member with relaxor (Rel) ferroelectrics such as PMN, the left side of the diagram before the M phase is occupied by the relaxor phase (figure 2.2) (Cox *et al* 2001). Thus, it is important to understand what drives the AFE-FE, R-T (MPB) and Rel-FE phase compositional phase transitions.

We first discuss the AFE-FE phase transition. Empirically, AFE in Pb-based systems is found when the B-site is occupied by large cations with small displacements such as Zr and $Yb_{0.5}Nb_{0.5}$ and then the phase transitions to FE as the B-cation size decreases to that of $Sc_{0.5}Nb_{0.5}$ and $Fe_{0.5}Nb_{0.5}$ (Isupov 2004, Chen 2000). These AFE structures exhibit O_6 tilts combined with essentially no displacements of the B-site and small antiparallel Pb displacements. Recent first-principles calculations shed light on this dependence of the relative stability of the AFE and FE phases on the B-cation size and displacement magnitude (Samanta *et al* 2022). Calculations of 180° domain wall structures showed that in the absence of A-cation displacements, the 180° domain wall structure with B-cation displacing in the opposite directions on the two sides of the domain wall was lower in energy than the corresponding single-domain structure where all B-cations were displaced in the same directions (while the A-site cation displacement was set to zero) (figure 2.9(a)). This result can be easily understood by considering that electrostatic interactions favor the creation of antiparallel chains of dipoles. By symmetry, the displacement of one cation sublattice (either A- or B-) in such antiparallel chains will force the other cation sublattice to be frozen at the high-symmetry positions and will incur an increased A–B repulsion energy cost. By contrast, in the ferroelectric phase, the

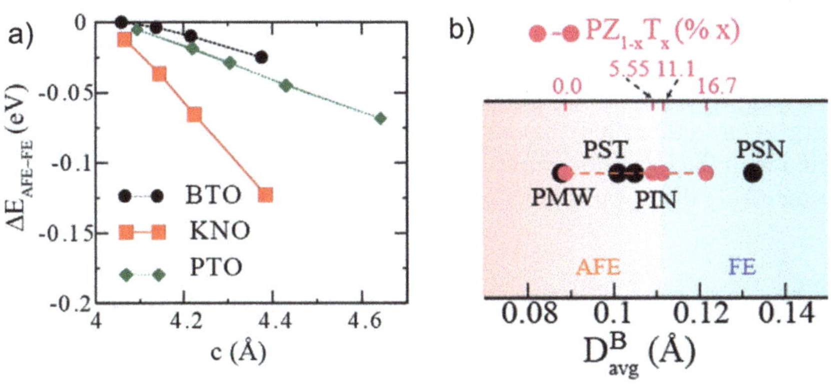

Figure 2.9. (a) Energy difference between the AFE and FE phase for the structures with the A-site frozen in the high-symmetry position as a function of lattice parameter for BTO ($BaTiO_3$), KNO ($KNbO_3$) and PTO ($PbTiO_3$) classic FE perovskites. (b) Phase stability of AFE and FE phases as a function of the average B-site displacement of the FE phase as obtained by DFT calculations. Reprinted from Samanta *et al* (2022). Copyright © 2022 Wiley.

displacements of all of the cations of one sublattice in the same direction allows the increased A–B repulsion energy cost to be alleviated by the displacements of the other cation sublattice. Thus, the AFE phase minimizes the electrostatic energy but maximizes the local A–B repulsion energy, while the opposite is the case for the FE phase. Therefore, for Pb-based perovskite oxides, small displacements of the B-cations will favor the AFE phase because the alleviation of the increased repulsion energy in the FE phase is quite weak due to the small displacement of the B-cation sublattice. For B-cations with large displacements such as Ti, the strong reduction in the A–B energy cost of the FE phase lowers its energy to below that of the AFE phase. Therefore, it can be expected that the AFE-FE phase stability of Pb-based perovskite oxides should be a function of the average magnitude of the B-cation displacements D_B^{avg}. As shown in figure 2.9(b), this is in fact observed for both $PbZr_{1-x}Ti_xO_3$ solid solutions and $Pb(\beta',\beta'')O_3$ solid solutions with AFE phase stable for $D_B^{avg} < 0.1$ and the FE phase stable for $D_B^{avg} > 0.1$. Thus, the consideration of the fundamental physical interactions in ABO_3 FE materials allows the identification of a simple and general criterion for the AFE-FE compositional phase transitions.

Next, we discuss the R–T compositional phase transition that gives rise to the MPB and high piezoelectric response. Comparison of the experimental phase diagrams for PZT and PMN-PT that have similar average intrinsic B-cation displacement magnitudes (0.13 Å for Zr and 0.16 Å for Mg/Nb) but different cation sizes (0.72 Å for Zr and 0.67 Å for $Mg_{1/3}Nb_{2/3}$) suggests that the MPB is shifted to lower Ti content by the smaller size of the $Mg_{1/3}Nb_{2/3}$ B-site of PMN compared to the larger size of the Zr B-site of PZT. In turn, comparison of PMN-PT and PZN-PT with essentially the same average B-cation size (0.68 Å for PZN and 0.67 Å for PMN) but with different average intrinsic average B-cation displacement magnitudes (0.20 Å for PZN and 0.16 Å for $Mg_{1/3}Nb_{2/3}$) shows that that the MPB is shifted to lower Ti content by the larger D_B^{avg} value of the $Zn_{1/3}Nb_{2/3}$ B-site. Consideration of the interplay between the structure and energetics of the R and T phases and the influence of the atomic R and D parameters on this interplay then allows the derivation of an analytical formula for predicting MBP location in terms of average *R* and *D* of the B-site (Grinberg *et al* 2005).

Experimental results (Corker *et al* 1998) and first-principles calculations (Grinberg *et al* 2002, 2004) have shown that the R phase in the Zr-rich PZT is disordered with the local Pb- and B-cation displacement vectors distributed in a cone around the (111) direction (figure 2.10(a)), while the T phase in the Ti-rich PZT is

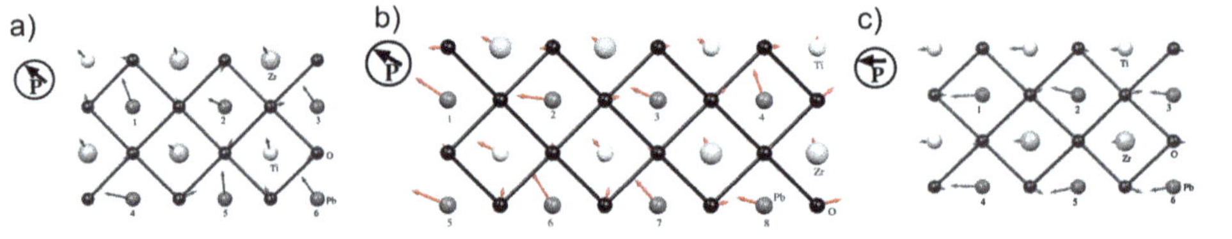

Figure 2.10. (a) DFT-relaxed structure of Zr-rich 30-atom PZT supercell. (b) DFT-relaxed structure of 50/50 40-atom PZT supercell. (c) DFT-relaxed structure of Ti-rich 30-atom PZT supercell. Reprinted from Grinberg *et al* (2004). Copyright © 2004 American Physical Society.

ordered with only small deviation of the local displacements vectors from the 100 direction (figure 2.10(c)) with intermediate behavior for the 50/50 M phase (figure 2.10(b)). It was also shown that in PZT, the Pb atoms tend to displace away from the larger Zr atoms and toward the smaller Ti atoms because Pb–Ti repulsion is weaker than Pb–Zr repulsion due to the smaller size of Ti compared to Zr (Grinberg *et al* 2002, 2004). At the same time, cation displacement scatter is energetically unfavorable due to the preference of the dipoles to align. Therefore, each ion displacement must make a trade-off between avoiding the larger Zr ions and achieving maximum alignment with the overall P direction. This trade-off is easy to achieve in the Ti-rich PZT due to the relatively small fraction of large Zr ions and therefore, all Pb atoms can displace along the 100 direction with only small deviations from the 100 direction. For the Zr-rich PZT, such a coordinated (100) displacement would incur a too large Pb–Zr repulsion energy cost; therefore, to reduce the Pb–Zr repulsive energy cost, the Pb displacements are along (111) on average (this reduces the A–B repulsion energy relative to the (100) displacement) with a large local scatter to avoid the Zr ions as much as possible (Grinberg *et al* 2002, 2004). Therefore, we can express the energy difference between the R and T phases in terms of the contributions of the electrostatic energy cost of dipole disorder ΔE_{dis} and the local repulsive A–B interactions $\Delta E_{\mathrm{A-B}}$ as

$$\Delta E_{\mathrm{T-R}}(x_{n\mathrm{PT}}) = -\Delta E_{\mathrm{dis}}(x_{\mathrm{nPT}}) + \Delta E_{\mathrm{A-B}}(x_{\mathrm{nPT}}) \tag{2.5}$$

where x_{nPT} is the fraction of the non-$PbTiO_3$ end-member of the solid solution (Grinberg *et al* 2005). For simplicity, ΔE_{dis} is assumed to be independent of composition because in solid solutions with $PbTiO_3$ the Pb displacements make the largest contribution to the dipole–dipole interactions and Pb displacement magnitudes vary weakly with the composition of the $PbTiO_3$-based solid solutions. On the other hand, due to the strong dependence of the local repulsions on the orbital overlap between the cations, $\Delta E_{\mathrm{A-B}}$ must show a strong composition dependence and is expressed as a function of average B-site ionic size $R_{\mathrm{B}}^{\mathrm{avg}}$ and average B-site intrinsic preferred displacement $D_{\mathrm{B}}^{\mathrm{avg}}$ to obtain

$$\Delta E_{\mathrm{A-B}}(x_{\mathrm{nPT}}) = x_{\mathrm{nPT}}(a_0 + b_0 R_{\mathrm{B}}^{\mathrm{avg}} - c_0 D_{\mathrm{B}}^{\mathrm{avg}}) + x_{\mathrm{PT}}(a_0 + b_0 R_{\mathrm{Ti}}^{\mathrm{avg}} - c_0 D_{\mathrm{Ti}}^{\mathrm{avg}}) \tag{2.6}$$

where x_{PT} is the fraction of $PbTiO_3$ in the solid solution. Here, greater $R_{\mathrm{B}}^{\mathrm{avg}}$ increases the preference for the R phase while greater $D_{\mathrm{B}}^{\mathrm{avg}}$ increases the preference for the T phase because greater $D_{\mathrm{B}}^{\mathrm{avg}}$ allows stronger displacement by the B-cation to decrease the A–B repulsion energy.

After some manipulation and fitting the constants to the experimental data for MPB locations, an expression for the $PbTiO_3$ content at the MPB in terms of $D_{\mathrm{B}}^{\mathrm{avg}}$ and $R_{\mathrm{B}}^{\mathrm{avg}}$ is obtained as

$$x_{\mathrm{PT}}^{\mathrm{MPB}} = 1 - 1/(-0.32 + 4.53 R_{\mathrm{B}}^{\mathrm{avg}} - 7.99 D_{\mathrm{B}}^{\mathrm{avg}}) \tag{2.7}$$

This expression shows a good agreement between predicted and actual MPB locations (figure 2.11). Equation (2.7) is clearly an approximation and in fact different values of the a_0, b_0, and c_0 constants must be used for solid solutions with

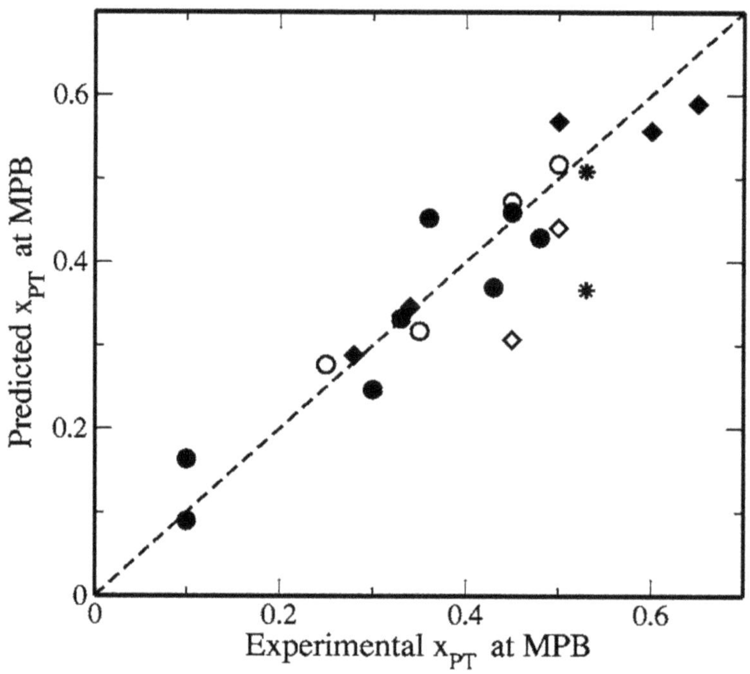

Figure 2.11. Experimental versus predicted MPB compositions obtained using equation (2.7). The solid solutions for which DFT B-cation displacement data are available are marked by filled cirles and diamonds for Pb-based and Bi-based systems, respectively. MPB positions predicted for Pb- and Bi-based systems where B-cation displacement data are estimated are marked by open circles and diamonds, respectively. Reprinted from Grinberg *et al* (2005). Copyright © 2005 AIP Publishing.

only Pb on the A-site and solid solutions with a mixture of Pb and Bi cations on the A-site. Nevertheless, equation (2.7) clearly captures the main trends and allows us to understand how the interatomic interactions and the average properties of constituent cations in solid solutions control the R–T transition and MPB location in ferroelectric solid solutions. Importantly, the accuracy of this expression has been confirmed for new experimentally synthesized materials including ternary, quaternary and quintary systems, underscoring the fact that despite the complexity of the solid solutions, their phase diagrams are controlled by the average properties of the A- and B-sites.

The third key compositional phase transition is the relaxor-to-ferroelectric phase transition. Unlike standard ferroelectrics, relaxors exhibit only local polarization at the nanoscale, with the nanodomain polarized regions separated by low-angle domain walls. The stability of the relaxor phase versus the standard FE phase is related to the variation in the local, Å-environment. This was demonstrated in 1980 in the classic work by Setter and Cross (Setter and Cross 1980) who showed that $PbSc_{0.5}Ta_{0.5}O_3$ (PST) with RS-ordered Sc/Ta B-cation arrangement is a standard ferroelectric, while the PST with disordered Sc/Ta B-cation arrangement is a relaxor showing diffuse and dispersive dielectric constant peak in the plot of the dielectric constant at different frequencies versus temperature (figure 2.12(a)). The difference

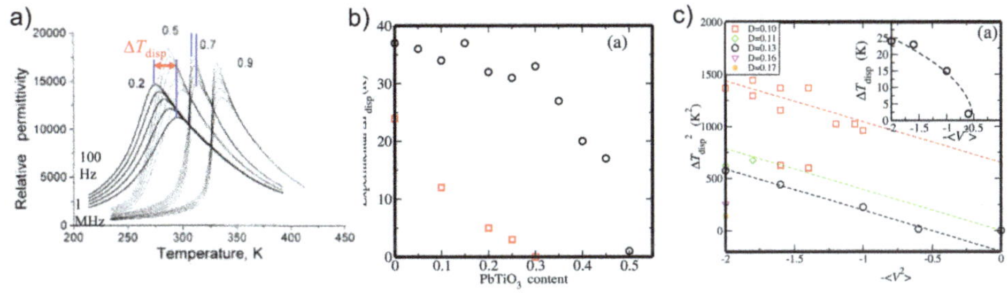

Figure 2.12. (a) Dielectric constant as a function of temperature at different frequencies and compositions for the PMN-PSN solid solutions. Reprinted from Farber and Davies (2003). Copyright © 2003 Wiley. (b) Dispersion as a function of $PbTiO_3$ content for PSW-PT (black) and PMN-PT (red) solid solutions. (c) Dispersion as a function of $\langle V^2 \rangle$ showing a clear relationship between valence variation, displacement and dispersion. (b), (c) Reprinted from Grinberg *et al* (2007). Copyright © 2007 American Physical Society.

between the temperature of the maximum dielectric constant at different frequencies $\Delta T_{\varepsilon,\max}$ (e.g. 100 Hz and 1 MHz) can be used as a measure of the strength of the relaxor behavior (Grinberg *et al* 2007). It is zero for the FE material and increases with greater compositional difference from the Rel-FE phase boundary (figure 2.12(b)).

Since a disordered Sc/Ta-cation arrangement in PST creates under- and over-bonded O atoms, it is natural to use the degree of O under- and overbonding as a measure of the variation of the local environment in these materials (Grinberg *et al* 2007). An underbonded (overbonded) O atom will attract (repel) Pb atoms to compensate the deficit (surplus) of B–O bonds. Thus, variation in the O–B–cation bond strengths is equivalent to the local variation in the potential experienced by the Pb atoms. For a Pb-based FE, the average B-cation neighbor valence of an O atom is 8, so that the variation in the local environment can be measured by the standard deviation of the total valence of the two B-cation neighbors of the O atoms. This can be mathematically expressed as

$$\langle V^2 \rangle = \frac{1}{N_{\mathrm{O}}} \sum_i \left(V_{i,1}^{\mathrm{B}} + V_{i,2}^{\mathrm{B}} - 2V_{\mathrm{avg}}^{\mathrm{B}}\right)^2 \tag{2.8}$$

where N_{O} is the total number of the O atoms in the unit cell, $V_{i,1}{}^{\mathrm{B}}$ and $V_{i,2}{}^{\mathrm{B}}$ are the valences of the two B-cation neighbors of the *i*th O atom, and $V_{\mathrm{avg}}{}^{\mathrm{B}}$ is the average B-cation valence that is equal to 4 in perovskite oxides with Pb atom on the A-site. For systems with long-range random-site order for the B-cation arrangement, the statistical distribution of the cations on the β' and β'' sites is known and therefore the probabilities of different B_1–O′–B_2 arrangements are also known. Therefore, $\langle V^2 \rangle$ can be evaluated exactly for these systems. Using the experimental data for the PMN-PT, $PbSc_{1/2}Ta_{1/2}O_3$–$PbTiO_3$ (PST-PT), PSW-PT, PZN-PT and PMN-PSN solid solutions with random-site ordering to plot $\Delta T_{\varepsilon,\max}$ versus $\langle V^2 \rangle$, a clear relationship can be observed (figure 2.12(c)). Furthermore, examination of the $D_{\mathrm{B}}{}^{\mathrm{avg}}$ of these solid solutions shows that the dependence of $\Delta T_{\varepsilon,\max}$ on $\langle V^2 \rangle$ is the same for all solid solutions, but the $\Delta T_{\varepsilon,\max}$ vs $\langle V^2 \rangle$ curves are shifted up for lower $D_B{}^{avg}$ and down for higher $D_B{}^{avg}$. This can be understood to be due to the greater ability of the B-cations with larger displacements to displace to alleviate O under- and over-bonding and thus decrease the variations in the local potential energy surface. Thus, similar to the R–T transition, the relaxor–FE compositional phase transition can be understood in terms of interatomic interactions and the average properties of the constituent atoms of ferroelectric solid solutions.

2.5 Ferroelectric-to-paraelectric transition and the Curie temperature

The ferroelectric-to-paraelectric (PE) transition temperature (the Curie temperature T_C) is a fundamental parameter of FE materials that sets the operating temperature range of piezoelectric and dielectric devices. There has been an ongoing strong interest in extending the temperature range of FE perovskite oxides to enable their use in high-temperature environments (Bell *et al* 2021). Additionally, it is empirically known that the coercive field that governs the switching of ferroelectric polarization is related to T_C, with higher coercive field obtained for materials with higher T_C values. Thus, it is important to understand the factors that govern T_C in complex perovskite oxide solutions.

Naively, it can be thought that since FE arises in part from the interactions between the individual dipoles created by cation off-center displacement, FE oxides with higher cation displacements and polarization values will have a higher T_C. This correlation was first observed by Abrahams *et al* (1968) who suggested that T_C is proportional to the displacement of the cation squared. Later, correlations between experimental T_C values and the square of DFT-calculated P at 0 K was demonstrated for Pb-based solid solutions and for $Bi(Me'Me'')O_3$–$PbTiO_3$ solid solutions (figure 2.13(a)) (Grinberg and Rappe 2004, 2007). This correlation can be explained by a simple model that considers the polarization of each individual unit cell to be fixed at all temperatures. Then, as the temperature is increased, the higher temperature results in the gradual entropy-driven disordering of the directions of the dipoles until at T_C the dipoles are oriented randomly so that the overall P is zero. In this Ising-model picture, the energy cost of dipole disorder is proportional to the strength of the interaction between the dipoles which in turn is proportional to the magnitude of the dipole squared due to the dependence of the electrostatic interaction on the magnitudes of both dipoles.

However, a closer inspection of the data for the different solid solutions of $PbTiO_3$ (figure 2.13(b)) shows (Yadav *et al* 2022) that in a plot of the experimental T_C versus P^2 the best fit is obtained for a line crossing the y-axis at a value greater than zero so that T_C is described by

$$T_C = c + \alpha P^2 \tag{2.9}$$

where c and α are constants. This is unphysical, because it suggests that a non-zero T_C will be obtained for a material with zero polarization. Rather, this result indicates that the relationship between T_C and P is more complicated than that

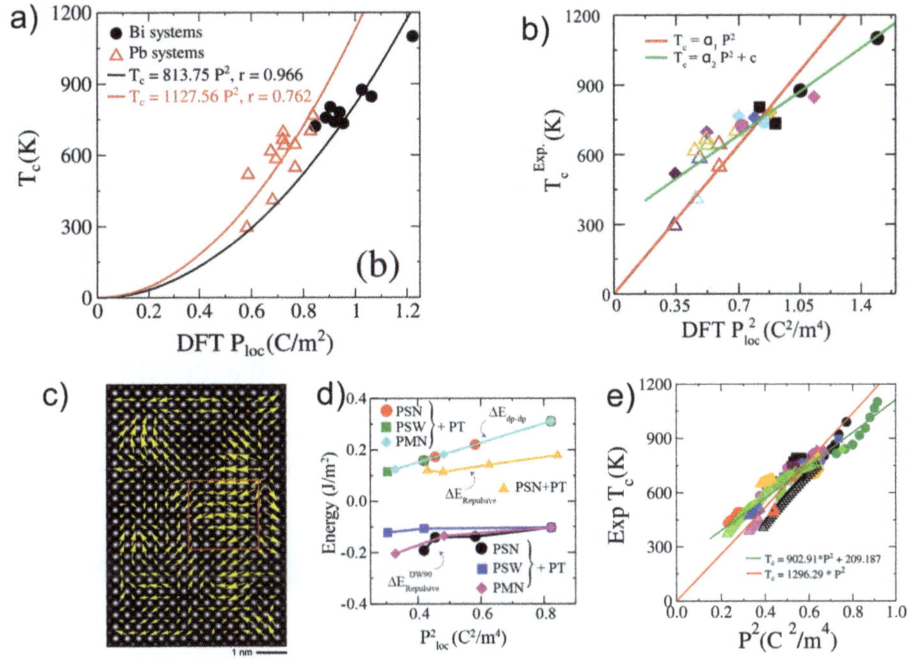

Figure 2.13. (a) Dependence of T_C on polarization for $PbTiO_3$ solid solutions with Pb only and mixed Bi/Pb occupation of the A-site. (b) Relationship between T_C and P^2. (c) Local dipole pattern in the paraelectric phase just above T_C in BST ($(Ba,Sr)TiO_3$). (d) Energy contribution to σ_{71} of different terms as a function of total P of the FE solid solution system. (e) Experimental T_C values plotted versus two different prediction equations. (a), (b), (d), (e) Reprinted from Yadav *et al* (2022). Copyright © 2022 Wiley. (c) Reprinted from Bencan *et al* (2021) Copyright © 2021 Springer Nature.

assumed by the Ising-model-like picture. Another curious fact is that there is no data for $P < 0.3$ C m^{-2} or $T_C < 200$ K, so that the relationship between T_C and P is in fact not known empirically for $PbTiO_3$-based solid solutions in this P and T_C range.

For a more accurate model of the FE-PE phase transition, it is necessary to take a closer look at the structure of the PE phase just above the phase transition. Simple energy considerations show that the lowest non-polar energy structure will not be fully disordered but rather will have regions of local order that avoid head-to-head or tail-to-tail dipole orientation but have different polarization orientation, leading to zero overall P. In recent work, such patterns have been observed experimentally by transmission electron microscopy (figure 2.13(c)) (Bencan *et al* 2021). Thus, the energy difference between the FE and PE phases should be related to the energy of the 71° R domain wall at 0 K, because it is this domain structure that shows local variation in P similar to that observed in the PE phase.

A similar concept is well-known in the studies of melting where it has been shown that the melting point of a solid is proportional to its vacancy formation energy rather than to its cohesive energy (Hayashiuchi *et al* 1982). This is because the liquid just above the melting point consists of clusters of atoms rather than of individual atoms.

Thus, to understand the empirically observed relationship between T_C and P^2, it is necessary to consider the energetics of 71° domain walls. Consideration of the 71° domain wall structure shows that the energy cost of the domain wall formation is due to the less favorable dipole–dipole interactions (dipole term) and due to the increased short-range repulsive interactions. As the polarization of the system becomes larger, the dipole term will increase as P^2. However, it was found that the short-range repulsive terms varies slowly and can be approximated by a constant for the $P > 0.3$ in $PbTiO_3$-based solid solutions (figure 2.13(d)). For smaller P values, this term will decay to zero but the functional form of the decay is unknown. Furthermore, no Pb-based perovskites with $P < 0.3$ are known, most likely because such small polarization can only be obtained for B-cations with small displacement magnitudes. As discussed above, such D_B^{avg} values will lead to the AFE phase being more stable than the FE phase. Thus, the T_C dependence of the complex Pb-based ferroelectric solutions can be easily understood based on the consideration of structure and energy cost of the local arrangement of dipoles in the PE phase just above the phase transition temperature (figure 2.13(e)).

2.6 Coercive field

The coercive field is a key parameter for devices based on FE oxides. For example, it sets the limit of applied power in transducers in which case higher fields are favored, and sets the energy cost of writing the FE memory in which case lower fields are favored. Switching in ceramic FE materials has been thought to be a complicated process that is sensitive to the details of the microstructure such as grain boundaries, surface and defects. Domain wall (DW) motion was known to be a process related to switching and since the early work of Merz (1956), it was known that the DW speed under applied field is proportional to $\exp(-E_{act}/E)$ where E_{act} is a parameter known as the activation field and E is the applied field. This relationship is known as Merz's law. Miller and Weinrech (MW) explained this functional form of the relationship between the DW motion speed and field by postulating that the DW motion is controlled by the nucleation at the DW (slow process) that is then followed by the (much faster) sideways and forward growth processes (Miller and Weinreich 1960). Elementary consideration of the competition between the volume energy term proportional to the electric field and the area of the nucleus and the interface energy term proportional to the DW energy and the perimeter of the nucleus naturally leads to the Merz's law dependence of the nucleation rate for the critical nucleus which is equivalent to the transition state in standard chemical kinetics. MW further assumed that the nucleus had a long triangular shape and was located at the surface of the FE material or its interface with the electrode. At the time, Miller and Weinreich did not have any means of estimating the DW energy and therefore derived the DW energy from the experimental data for E_{act}, obtaining the 180° DW energy σ_{180} of ~0.6 mJ m^{-2} for $BaTiO_3$. While the MW model was widely accepted, four decades later, following the calculation of DW energy values by first-principles methods, experimental measurements of DW motion speed in PZT by Tybell *et al* (2002) found that the observed DW motion speed was inconsistent with the MW model. Using the DFT-calculated σ_{180}

value for $PbTiO_3$ (Meyer and Vanderbilt 2002) results in the estimate of the critical nucleus size that is larger than the film thickness, suggesting very slow DW motion inconsistent with the fast DW dynamics observed experimentally. Therefore, Tybell *et al* suggested that effects other than standard nucleation at the DW were responsible for DW motion or alternatively, DW motion was promoted by defects. Similarly, comparison of the σ_{180} estimated by MW for $BaTiO_3$ to the σ_{180} value for $BaTiO_3$ calculated by Meyer and Vanderbilt (2002) shows that the DFT value is larger by an order of magnitude than the experimentally-derived MW estimate. This pointed to the need to reconsider the classic MW model of DW motion.

As described in detail in chapter 3, this problem was resolved by multi-scale simulations of DW motion and P switching, which showed that while E_c (coercive field) is largely controlled by the nucleation at the DW in a perfect material, the nucleus shape and the interface energy cost are different from those assumed by Miller and Weinreich. However, similar to the original MW model, the critical nucleus energy (transition state energy) is proportional to σ^2. This naturally suggests that despite the complexity of the switching process, the dependence of E_c on the composition of complex ferroelectric systems can be understood and this understanding can be encapsulated in an analytical formula by exploiting the general relationship between the DW energy and P described above.

For bulk rhombohedral ferroelectric perovskite oxides, the switching is driven by the motion of 71° domain walls and the E_c measured at a temperature T [$E_c(T)$] scales with $\sigma_{71}(T)^2/P(T)$ (Shin *et al* 2007, Liu *et al* 2016); since σ_{71} is largely due to dipole–dipole interactions and scales with P^2, it is obtained that $E_c(T) \sim P^3(T)$. Using the dependence of P on T according to simple Landau theory $P(T) = (P_0^2/T_C)(T - T_C)^{1/2}$, where P_0 is the 0 K polarization and the above described relationship between P and T_C, it is obtained that $E_c(300\text{ K}) \sim (T_C - 300)^{1.5}$. As shown in figure 2.14(a), this formula describes well the relationship between the experimental results for T_C and for E_c measured at room temperature for single-crystal ferroelectric perovskite oxide solid solution compositions. For tetragonal perovskite ferroelectrics, the switching is driven by the motion of 90° T DWs governed by the energy of the 90° T DW σ_{90}, and a similar relationship can be derived by using T-phase σ_{90} instead of R-phase σ_{71}. Since for the same (or similar) composition, the tetragonal 90° DW energy σ_{90} is higher than the rhombohedral 71° DW energy σ_{71},

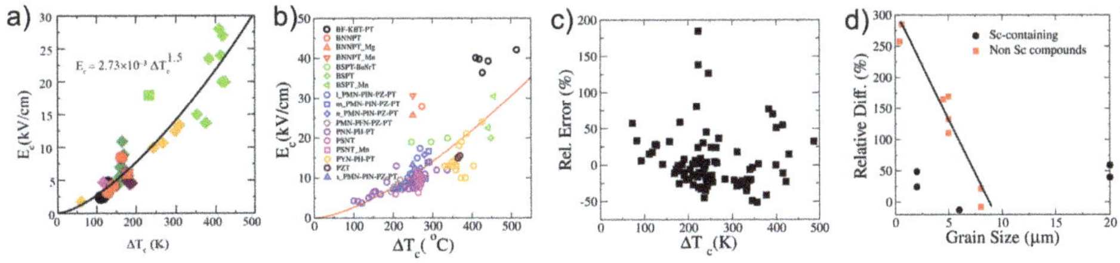

Figure 2.14. (a) E_c at 300 K plotted versus the temperature difference from T_C for single crystals. The line shows a fit to $E_c = (T_C - 300\text{ K})^{1.5}$. (b) Same as (a) but for ceramics. (c) Ratio of ceramic E_c to E_c predicted by fit to the single-crystal data versus ΔT_C. (d) Plot of the ratio of ceramic E_c to E_c predicted by fit to the single-crystal data versus grain size. Reprinted from Yadav *et al* (2022). Copyright © 2022 Wiley.

the coercive fields of T-phase ferroelectrics are higher as observed experimentally (Suchomel 2005).

For ceramic materials, the intrinsic coercive field predicted by $E_c(300\text{ K}) \sim (T_C - 300)^{1.5}$ can be used to evaluate the effect of the microstructure on switching by comparing to the actual measured coercive field of the ceramics (Yadav *et al* 2022). It was shown that the E_c values experimentally observed for ceramics can be both lower and higher than the intrinsic E_c values estimated by the proportionality to T_C, with strong E_c enhancement and weak E_c suppression. It was further shown that ceramics with larger grain size have E_c values closer to the intrinsic E_c estimate and that the deviation of the ceramic E_c from the intrinsic E_c value can be used to measure the degree of texture in small-grain ceramics (figures 2.14(b)–(d)). Thus, the understanding of trends in P, T_C and interatomic interactions at the DW can provide baseline information using simple formulas for the study of switching in complex ferroelectric solutions.

2.7 Perspective and future directions

As described above, theoretical research efforts over the course of the last three decades have been able to rationalize and explain the rich variety of behaviors observed in complex ferroelectric solid solutions based on fundamental interactions between the constituent atoms. It was demonstrated that for several key properties, the average characteristics of the solid solution and their degree of variation can quantitatively predict the behavior of the material. Thus, complex systems follow simple understandable trends. The success of this research effort for solid solutions of $PbTiO_3$ provides a firm basis for the application of this approach to more difficult problems. For example, lead-free ferroelectrics are currently under intense investigation due to the interest in replacing the lead-based PZT in applications (Koruza *et al* 2018, Hong *et al* 2016, Wu 2020). Development of composition–structure–property correlations and guiding principles for the design of high-performance materials with a full set of desired properties is difficult for these systems due to the smaller energy scales for the different phases and off-center displacements compared to the Pb-based systems, and to the even-higher complexity of the high-performance solid solutions discovered to date (Koruza *et al* 2018, Hong *et al* 2016, Wu 2020).

Another difficult task is to understand what controls the response properties of complex ferroelectric solid solutions such as the piezoelectric coefficients. For ceramics, this task is complicated by the strong effect of the microstructure on, e.g. the piezoelectric coefficient d_{33}. Thus, it will be necessary to understand how the composition of a solid solution affects its microstructure characteristics such as grain size and texture. Additionally, it has been long-known that doping (<2%) of the solid solution can strongly affect its d_{33}, and in particular, recent work has shown that Sm doping can lead to increases in d_{33} by a factor of two or more (Li *et al* 2018, 2019). Since doping is an extrinsic effect, the modeling and understanding of doping effects in terms of the composition and local structure of the material is more challenging. Finally, the obtained theoretical understanding should be applied to the design of practical new materials.

References

Abrahams S C, Kurtz S K and Jamieson P B 1968 Atomic displacement relationship to Curie temperature and spontaneous polarization in displacive ferroelectrics *Phys. Rev.* **172** 551

Akbas M A and Davies P K 1997 Domain growth in Pb $(Mg_{1/3}Ta_{2/3})O_3$ perovskite relaxor ferroelectric oxides *J. Am. Ceram. Soc.* **80** 2933–6

Bell A J, Comyn T P and Stevenson T J 2021 Expanding the application space for piezoelectric materials *APL Mater.* **9** 010901

Bencan A, Oveisi E, Hashemizadeh S, Veerapandiyan V K, Hoshina T, Rojac T, Deluca M, Drazic G and Damjanovic D 2021 Atomic scale symmetry and polar nanoclusters in the paraelectric phase of ferroelectric materials *Nat. Commun.* **12** 1–9

Bersuker I B 2018 Vibronic (pseudo Jahn–Teller) theory of ferroelectricity: novel aspects and applications *Ferroelectrics* **536** 1–59

Brown I D 2009 Recent developments in the methods and applications of the bond valence model *Chem. Rev.* **109** 6858–919

Burns G and Dacol F H 1983 Glassy polarization behavior in ferroelectric compounds Pb $(Mg_{1/3}Nb_{2/3})O_3$ and Pb $(Zn_{1/3}\ Nb_{2/3})O_3$ *Solid State Commun.* **48** 853–6

Burton B P and Cockayne E 1999 Why Pb (B, B′)O_3 perovskites disorder at lower temperatures than Ba (B, B′)O_3 perovskites *Phys. Rev.* B **60** R12542

Chen I W 2000 Structural origin of relaxor ferroelectrics—revisited *J. Phys. Chem. Solids* **61** 197–208

Cohen R E 1992 Origin of ferroelectricity in perovskite oxides *Nature* **358** 136–8

Corker D L, Glazer A M, Whatmore R W, Stallard A and Fauth F 1998 A neutron diffraction investigation into the rhombohedral phases of the perovskite series *J. Phys. Cond. Matt.* **10** 6251

Cox D E, Noheda B, Shirane G, Uesu Y, Fujishiro K and Yamada Y 2001 Universal phase diagram for high-piezoelectric perovskite systems *Appl. Phys. Lett.* **79** 400–2

Davies P K and Akbas M A 2000 Chemical order in PMN-related relaxors: structure, stability, modification, and impact on properties *J. Phys. Chem. Solid* **61** 159–66

Davies P K 1999 Cation ordering in complex oxides *Curr. Op. Solid State and Mater. Sci.* **4** 467–71

Davies P K, Wu H, Borisevich A Y, Molodetsky I E and Farber L 2008 Crystal chemistry of complex perovskites: new cation-ordered dielectric oxides *Annu. Rev. Mater. Res.* **38** 369–401

Du X H, Zheng J, Belegundu U and Uchino K 1998 Crystal orientation dependence of piezoelectric properties of lead zirconate titanate near the morphotropic phase boundary *Appl. Phys. Lett.* **72** 2421–3

Eng H W, Barnes P W, Auer B M and Woodward P M 2003 Investigations of the electronic structure of d0 transition metal oxides belonging to the perovskite family *J. Solid State Chem.* **175** 94–109

Eremenko M, Krayzman V, Bosak A, Playford H Y, Chapman K W, Woicik J C, Ravel B and Levin I 2019 Local atomic order and hierarchical polar nanoregions in a classical relaxor ferroelectric *Nature Commun.* **10** 1–9

Farber L and Davies P K 2003 Influence of cation order on the dielectric properties of Pb $(Mg_{1/3}\ Nb_{2/3})O_3$–Pb $(Sc_{1/2}\ Nb_{1/2})O_3$ (PMN–PSN) relaxor ferroelectrics *J. Am. Ceram. Soc.* **86** 1861–6

Fu D, Taniguchi H, Itoh M, Koshihara S Y, Yamamoto N and Mori S 2009 Relaxor Pb $(Mg_{1/3}Nb_{2/3})O_3$: a ferroelectric with multiple inhomogeneities *Phys. Rev. Lett.* **103** 207601

Goldschmidt V M 1926 Goldshmidt tolerance factor *Die Nature Wissen.* **14** 477–85

Grinberg I and Rappe A M 2004 Local structure and macroscopic properties in $PbMg_{1/3}Nb_{2/3}O_3$–$PbTiO_3$ and $PbZn_{1/3}Nb_{2/3}O_3$–$PbTiO_3$ solid solutions *Phys. Rev.* B **70** 220101

Grinberg I and Rappe A M 2007 Nonmonotonic T_c trends in Bi-based ferroelectric perovskite solid solutions *Phys. Rev. Lett.* **98** 037603

Grinberg I, Cooper V R and Rappe A M 2002 Relationship between local structure and phase transitions of a disordered solid solution *Nature* **419** 909–11

Grinberg I, Cooper V R and Rappe A M 2004 Oxide chemistry and local structure of $PbZr_xTi_{1-x}O_3$ studied by density-functional theory supercell calculations *Phys. Rev.* B **69** 144118

Grinberg I, Juhás P, Davies P K and Rappe A M 2007 Relationship between local structure and relaxor behavior in perovskite oxides *Phys. Rev. Lett.* **99** 267603

Grinberg I, Suchomel M R, Davies P K and Rappe A M 2005 Predicting morphotropic phase boundary locations and transition temperatures in Pb- and Bi-based perovskite solid solutions from crystal chemical data and first-principles calculations *J. Appl. Phys.* **98** 094111

Guo R, Cross L E, Park S E, Noheda B, Cox D E and Shirane G 2000 Origin of the high piezoelectric response in $PbZr_{1-x}Ti_xO_3$ *Phys. Rev. Lett.* **84** 5423

Hayashiuchi Y, Hagihara T and Okada T 1982 A new interpretation of proportionality between vacancy formation energy and melting point *Physica* B + C **115** 67–71

Hong C H, Kim H P, Choi B Y, Han H S, Son J S, Ahn C W and Jo W 2016 Lead-free piezoceramics–where to move on? *J. Materiom.* **2** 1–24

Isupov V A 2004 Problems in crystal chemistry of ferroelectric and antiferroelectric perovskites $PbB'_{0.5}\,B''_{0.5}O_3$ *Crystallog. Rep.* **49** 719–23

Jaffe B, Roth R S and Marzullo S 1954 Piezoelectric properties of lead zirconate-lead titanate solid-solution ceramics *J. Appl. Phys.* **25** 809–10

Juhás P, Davies P K and Akbas M A 2004a Structure and dielectric properties of $Pb(S_{2/3}\,W_{1/3})O_3$–$Pb(Zr/Ti)O_3$ relaxors *J. Am. Ceram. Soc.* **87** 2086–92

Juhas P, Grinberg I, Rappe A M, Dmowski W, Egami T and Davies P K 2004b Correlations between the structure and dielectric properties of $Pb(Sc_{2/3}\,W_{1/3})O_3$–$Pb(Ti/Zr)O_3$ relaxors *Phys. Rev.* B **69** 214101

Khalyavin D D, Salak A N, Vyshatko N P, Lopes A B, Olekhnovich N M, Pushkarev A V, Maroz I I and Radyush Y V 2006 Crystal structure of metastable perovskite Bi $(Mg_{1/2}\,Ti_{1/2})O_3$: Bi-based structural analogue of antiferroelectric $PbZrO_3$ *Chem. Mater.* **18** 5104–10

Koruza J, Bell A J, Froemling T, Webber K G, Wang K and Rödel J 2018 Requirements for the transfer of lead-free piezoceramics into application *J. Materiom.* **4** 13–26

Laurita G and Seshadri R 2022 Chemistry, structure, and function of lone pairs in extended solids *Acc. Chem. Res.* **55** 1004–14

Li F *et al* 2019 Giant piezoelectricity of Sm-doped Pb $(Mg_{1/3}\,Nb_{2/3})O_3$–$PbTiO_3$ single crystals *Science* **364** 264–68

Li F *et al* 2018 Ultrahigh piezoelectricity in ferroelectric ceramics by design *Nature Mater.* **17** 349–54

Liu S, Grinberg I and Rappe A M 2016 Intrinsic ferroelectric switching from first principles *Nature* **534** 360

Lu X, Zhang H, Zheng L and Cao W 2016 Phase coexistence in ferroelectric solid solutions: formation of monoclinic phase with enhanced piezoelectricity *AIP Adv.* **6** 105208

Luo L, Zhao X and Luo H 2010 Single crystal PZN-PT, PMN-PT, PSN-PT, and PIN-PT-based piezoelectric materials *Adv. Piezo. Mater.* 271–318

Merz W J 1956 Switching time in ferroelectric $BaTiO_3$ and its dependence on crystal thickness *J. Appl. Phys.* **27** 938–43

Meyer B and Vanderbilt D 2002 *Ab initio* study of ferroelectric domain walls in $PbTiO_3$ *Phys. Rev.* B **65** 104111

Miller R C and Weinreich G 1960 Mechanism for the sidewise motion of 180 domain walls in barium titanate *Phys. Rev.* **117** 1460

Noheda B, Cox D E, Shirane G, Gonzalo J A, Cross L E and Park S E 1999 A monoclinic ferroelectric phase in the $Pb(Zr_{1-x}Ti_x)O_3$ solid solution *Appl. Phys. Lett.* **74** 2059–61

Noheda B, Cox D E, Shirane G, Park S E, Cross L E and Zhong Z 2001 Polarization rotation via a monoclinic phase in the piezoelectric 92% $PbZn_{1/3}Nb_{2/3}O_3$–8% $PbTiO_3$ *Phys. Rev. Lett.* **86** 3891

Polinger V, Garcia-Fernandez P and Bersuker I B 2015 Pseudo Jahn–Teller origin of ferroelectric instability in $BaTiO_3$ type perovskites: the green's function approach and beyond *Physica* B **457** 296–309

Samanta A *et al* 2022 A predictive theory for domain walls in oxide ferroelectrics based on interatomic interactions and its implications for collective material properties *Adv. Mater.* **34** 2106021

Setter N and Cross L E 1980 The role of B-site cation disorder in diffuse phase transition behavior of perovskite ferroelectrics *J. Appl. Phys.* **51** 4356–60

Shin Y-H, Grinberg I, Chen I-W and Rappe A M 2007 Nucleation and growth mechanism of ferroelectric domain-wall motion *Nature* **449** 881

Shirane G, Hoshino S and Suzuki K 1950 Crystal structure of lead titanate and of lead-barium titanate *J. Phys. Soc. Jap.* **5** 453–55

Suchomel M R 2005 *Greater Functionality of Bismuth and Lead-based Perovskites* (Philadelphia, PA: University of Pennsylvania)

Takenaka H, Grinberg I, Liu S and Rappe A M 2017 Slush-like polar structures in single-crystal relaxors *Nature* **546** 391–5

Tian J, Han P, Huang X, Pan H, Carroll J F and Payne D A 2007 Improved stability for piezoelectric crystals grown in the lead indium niobate–lead magnesium niobate–lead titanate system *Appl. Phys. Lett.* **91** 222903

Tsukada S, Ohwada K, Ohwa H, Mori S, Kojima S, Yasuda N, Terauchi H and Akishige Y 2017 Relation between fractal inhomogeneity and In/Nb-arrangement in Pb $(In_{1/2}Nb_{1/2})O_3$ *Sci. Rep.* **7** 1–8

Tybell T, Paruch P, Giamarchi T and Triscone J M 2002 Domain wall creep in epitaxial ferroelectric $Pb(Zr_{0.2}Ti_{0.8})O_3$ thin films *Phys. Rev. Lett.* **89** 097601

von Hippel A 1950 Ferroelectricity, domain structure, and phase transitions of barium titanate *Rev. Mod. Phys.* **22** 221

Wheeler R A, Whangbo M H, Hughbanks T, Hoffmann R, Burdett J K and Albright T A 1986 Symmetric vs asymmetric linear MXM linkages in molecules, polymers, and extended networks *J. Am. Chem. Soc.* **108** 2222–36

Wu J 2020 Perovskite lead-free piezoelectric ceramics *J. Appl. Phys.* **127** 190901

Yadav S, Samanta A, Shafir O and Grinberg I 2022 A multilevel analytical theory for prediction of ferroelectric perovskite oxide properties from composition *Adv. Mater.* **34** 2106105

Ye Z G, Noheda B, Dong M, Cox D and Shirane G 2001 Monoclinic phase in the relaxor-based piezoelectric/ferroelectric $Pb(Mg_{1/3}Nb_{2/3})O_3$–$PbTiO_3$ system *Phys. Rev.* B **64** 184114

Zhang S, Luo J, Hackenberger W and Shrout T R 2008 Characterization of Pb $(In_{1/2}Nb_{1/2})$ O_3–Pb $(Mg_{1/3}Nb_{2/3})$ O_3–$PbTiO_3$ ferroelectric crystal with enhanced phase transition temperatures *J. Appl. Phys.* **104** 064106

Chapter 3

Multiscale simulations of ferroelectric oxides

Jiawei Huang, Zhuang Qian, Yihao Hu, Qisheng Yu and Shi Liu

The functionalities of ferroelectric materials often depend on the complex interactions between structures and processes spanning over multiple time and length scales. First-principles calculations, molecular dynamics, and phase field modeling have each contributed to the understandings of ferroelectrics at different levels. A multiscale approach that progressively bridges these methods will bring new insights into the rich emergent phenomena afforded by ferroelectrics. Machine learning methodologies will serve as a key component in the multiscale approach that may eventually enable facile computational design of complex multicomponent ferroelectrics.

3.1 Rationale of a multiscale approach

In 1972, P W Anderson rightfully stated in his now famous paper, 'More is different' (Anderson 1972), that complex physical systems may possess emergent properties that cannot be understood only in terms of the laws deduced from a microscopic picture. The Nobel Prize in Physics 2021 focused on the complexity of physical systems and the efforts made to achieve predictability. Much like Earth's climate and spin glass, ferroelectric materials can also be considered as a complex system, and their functionalities are dictated by the emergent properties. Even for an ideal single-crystal ferroelectric, the polarization switching process consists of a series of hierarchically organized physical processes spanning multiple time scales: from dipole flipping to domain nucleation and growth, and to domain wall motion. The switching becomes dramatically more complicated in polycrystalline ferroelectrics due to the inevitable chemical inhomogeneity on length scales ranging from angstroms and nanometers (e.g. defects and dislocations) to micrometers (e.g. grain boundaries). It is expected that the macroscopic properties of ferroelectrics associated with switching dynamics will depend on a network of processes across multiple scales in space and time. A fundamental challenge for rational design and optimization of ferroelectrics is thus to predict the response of ferroelectrics to external stimuli taking into account various intimately coupled intrinsic and extrinsic effects. Addressing this challenge is

doi:10.1088/978-0-7503-3975-9ch3

beyond the capability of a single theory or method. Therefore, a multiscale approach that progressively bridges methods at different scales is a natural choice.

Three levels of methods are widely adopted to model ferroelectrics: first-principles density functional theory (DFT), molecular dynamics (MD) simulations, and phenomenological phase-field model (PhFM). Using the positions of atoms as the only inputs, the DFT method allows for accurate predictions of various important material properties of ferroelectrics such as lattice constants, remnant polarization, and dielectric, piezoelectric, and pyroelectric coefficients (de Gironcoli *et al* 1989, Sághi-Szabó *et al* 1998, Bellaiche *et al* 2000, Liu *et al* 2016a, Tan *et al* 2018). However, first-principles DFT calculations are rather demanding computationally, and are only capable of handling small systems of a few hundred atoms. For this reason, the study of dynamical properties of ferroelectrics remains difficult for conventional DFT methods. Classical MD simulations is an ideal technique for investigating dynamical processes on larger time and length scales while providing atomistic details with femtosecond time resolution. The efficiency of classical MD is due to the usage of a classical force field that approximates the atomic interactions with a set of relatively simple analytical functions. The accuracy of the force field thus determines the accuracy of MD simulations. Developing a force field is often a tedious process, particularly for transition metal oxides. The limited availability and accuracy of classical force fields has been a bottleneck for the applications of MD simulations. To model ferroelectrics at the mesoscopic length scale, PhFM is the most popular and powerful method. Nevertheless, the PhFM method is subjected to three well-known limitations: (i) the material coefficients involved in the free energy expansion are difficult to determine, largely relying on fitting to known experimental data; (ii) the results of PhFM (e.g. phase diagram) may be sensitive to small variations of material coefficients; (iii) such phenomenological models cannot reveal atomistic mechanisms. A multiscale approach can overcome the limitations of each method and strike a balance between accuracy and efficiency. A classical force field for MD can be obtained by fitting to a database of DFT results to combine the efficiency of MD and the accuracy of DFT (Stefano *et al* 2012, Jain *et al* 2013, Garrity *et al* 2014, de Jong *et al* 2015). Using DFT and MD calculations, it is possible to further parametrize a Ginzburg–Landau-type free energy functional for PhFM (Völker *et al* 2011). Within such a multiscale approach, there are two central challenges: (i) how to faithfully represent complex and highly nonlinear potential energy surface with a classical force field for MD simulations? (ii) How to faithfully link microscopic variables at fine scales to governing equations at coarser scales? This chapter will show that machine learning (ML) methods offer opportunities to address both challenges.

In this chapter, a few typical atomistic potentials of ferroelectrics for MD simulations, including an ML-based potential, are firstly introduced. Recent applications of PhFM to study topological polar structures in ferroelectric oxide thin films are briefly reviewed. Pertinent to the computation-guided design of ferroelectric solid solutions, a few examples of data-driven investigations that utilize ML methods for accelerated discovery and characterization of complex ferroelectrics are discussed. Subsequently, the chapter describes a progressive theoretical framework that combines DFT, MD, and PhFM, which eventually allows for an efficient and accurate estimation of coercive fields for a broad range of ferroelectrics. Finally,the challenges and opportunities for multiscale simulations of ferroelectric materials are discussed.

3.2 Molecular dynamics

In classical MD simulations, all atoms, treated as structureless classical particles, interact through a classical force field and evolve following Newton's laws of motion during which the time-dependent behavior of an ensemble of atoms can be simulated efficiently (Metropolis *et al* 1953, Alder and Wainwright 1959, Rahman 1964, Smit and Frenkel 2001). It is evident that a force field which accurately describes the interatomic potential is the prerequisite of running an MD simulation. The conventional force field development is to approximate the interatomic potential with a set of analytical functions. In practice, the analytical energy function often includes two-body and many-body terms, long-range and short-range terms, and electrostatic and non-electrostatic terms. Once the form of the energy function is determined, the force field parameters are obtained through fitting such that MD simulations can reproduce known reference data. However, there is no guaranty that an energy function of 'ad hoc' forms with a limited number of parameters could faithfully represent the 'true' interatomic potential that is intrinsically high-dimensional. Consequently, developing an accurate force field for a ferroelectric is often a tedious process taking multiple months or even years. Here, the emphases are put on two representative force fields of ferroelectrics, shell model and bond valence model. Following this, the latest progress in force field development based on ML methods that open up exciting opportunities to develop transferable and expandable force fields with accuracy comparable with DFT, will be presented.

3.2.1 Shell model

In the shell model (Mitchell and Fincham 1993, Sepliarsky *et al* 2005), each atom is represented in terms of two coupled components, a positively charged core and a negatively charged shell. The core and shell are linked by a harmonic or anharmonic spring. Shells can be massless or possess small mass: the former will respond instantaneously to the motions of cores while the latter evolves according to Newton's laws of motion. In practice, shells with small mass have been proved to be more efficient computationally (Mitchell and Fincham 1993). By fitting to first-principles results, the shell model parameters for a few ferroelectric perovskites have been developed, including $PbTiO_3$ (Sepliarsky and Cohen 2002), $BaTiO_3$ (Tinte *et al* 1999, Vielma and Schneider 2013), $KNbO_3$ (Sepliarsky *et al* 1995), $Ba(Ti,Sr)O_3$ (Tinte *et al* 2004), $Pb(Zr,Ti)O_3$ (Gindele *et al* 2015), and $(1-x)Pb(Mg_{1/3}\,Nb_{2/3}\,O_3)$-$xPbTiO_3$ (Sepliarsky and Cohen 2011). Figure 3.1 schematically illustrates the interactions between two atoms in the shell model which contains three types of interactions. The interaction between the core and the shell of the same atom (V_{CS}) is described as an isotropic, nonlinear spring in the form of

$$V_{CS}(r) = \frac{1}{2}k_2r^2 + \frac{1}{24}k_4r^4, \tag{3.1}$$

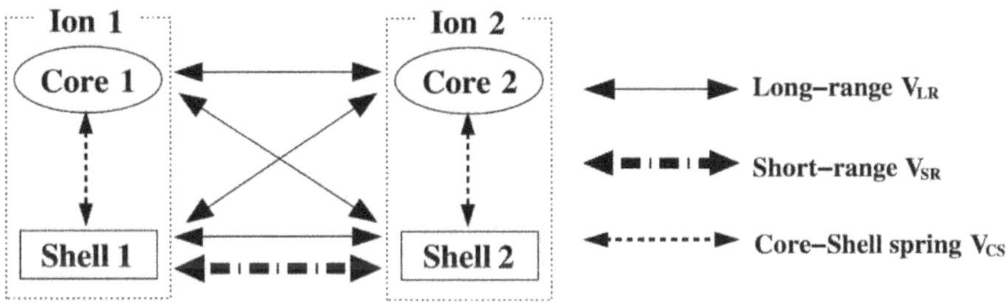

Figure 3.1. Schematic illustration of pair potential interactions between cores and shells in the framework of the shell model. Reprinted from (Shimada *et al* 2008). Copyright © 2008 IOP Publishing.

where k_2 and k_4 are spring constants and r is the distance between the core and the shell. The short-range interaction (V_{SR}) that captures the Pauli repulsion and van der Waals attraction between shells is modeled using the Buckingham potential,

$$V_{SR}(r) = \left\{ A \exp\left(-\frac{r}{\rho}\right) - \frac{c}{r^6} \right\} f_c(r), \tag{3.2}$$

where A, ρ, and c are potential parameters, and $f_c(r)$ is a smooth function that ensures the potential decays to zero smoothly from R_{c1} to R_{c2}.

$$f_c(r) = \begin{cases} 1, & (r \leqslant R_{c1}) \\ \frac{1}{2}\cos\left(\frac{r - r_{c1}}{r_{c2} - r_{c1}}\pi\right) + \frac{1}{2}, & (R_{c1} < r \leqslant R_{c2}) \\ 0, & (R_{c2} < r) \end{cases} \tag{3.3}$$

The long-range Coulomb potential (V_{LR}) has a simple form,

$$V_{LR} = \frac{1}{4\pi\varepsilon_0}\frac{q_i q_j}{r}, \tag{3.4}$$

where q is the particle charge and ε_0 is the vacuum permittivity.

The shell model has been proved to be able to successfully predict the static properties such as lattice parameters, phonon dispersion relationships as well as dynamical properties such as phase transition temperatures and domain nucleation and switching process. For example, the temperature-driven phase transitions have been simulated with the shell model for many different ferroelectrics. As shown in figure 3.2, early in 2000, Sepliarsky *et al* studied the phase diagram of $KNbO_3$ using a shell model (Sepliarsky *et al* 2000) and reproduced three successive temperature-driven phase transitions occurring at 225 K, 475 K, and 675 K, which are very close to experimental values of 210 K, 488 K, and 701 K, respectively. Vielma and Schneider (2013) reproduced the correct sequence of phase transitions of $BaTiO_3$: rhombohedral (180 K) → orthorhombic (250 K) → tetragonal (340 K) → cubic, compared to the experimental phase transitions at 183 K, 278 K, and 393 K, respectively. Similarly accurate predictions on temperature-driven phase transitions have also been made for $BiFeO_3$ (Graf *et al* 2014). The shell model gives a spontaneous polarization $P_s = 77$ μC cm^{-2}, comparable with the value of 60–100 μC cm^{-2} reported

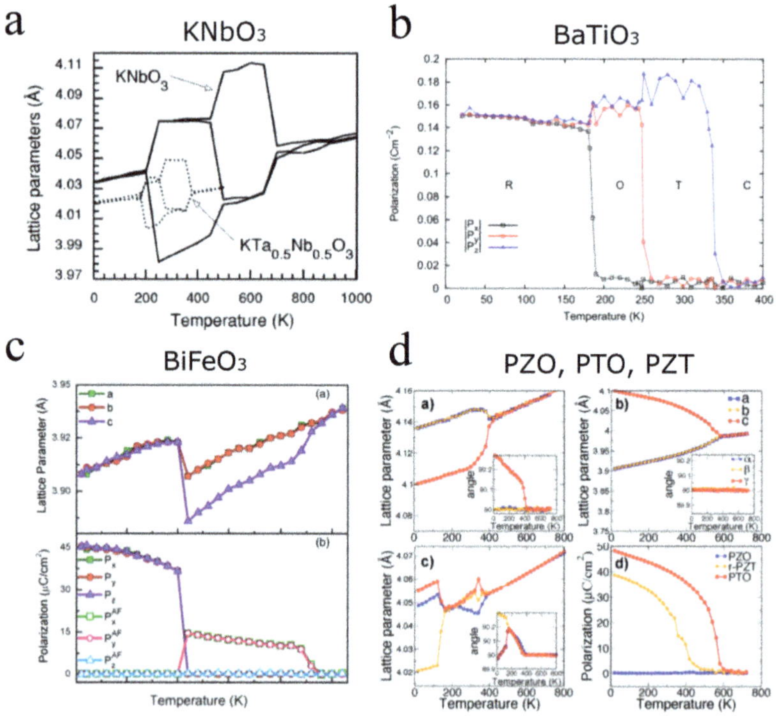

Figure 3.2. (a) Phase diagram of $KNbO_3$ (solid lines), determined from MD simulations using a shell model. The phase diagram of $KTa_{0.5}Nb_{0.5}O_3$ (dashed lines) is qualitatively identical to that of $KNbO_3$. Reprinted from (Sepliarsky *et al* 2000). Copyright © 2000 AIP Publishing LLC. (b) Average absolute polarization along all three Cartesian axes for $BaTiO_3$ as a function of temperature determined from MD simulations using a shell model fitted to scaled energy differences. Reprinted from (Vielma and Schneider 2013). Copyright © 2013 AIP Publishing LLC. (c) Phase diagram of $BiFeO_3$ at zero pressure resulting from the shell model MD simulations. Thermal evolution of the lattice parameters, net polarization, P_{net}, and order parameter associated with the antiferroelectric ordering along the z direction, P^{AF}. Reprinted from (Graf *et al* 2014). Copyright © 2014 American Physical Society. (d) Temperature dependence of the lattice parameters and the magnitude of polarization for Pb(Zr, Ti)O_3 solid solutions. Reprinted from (Gindele *et al* 2015). Copyright © 2015 American Chemical Society.

for single crystals at room temperature (Lebeugle *et al* 2007). The shell model exhibits good transferability, as demonstrated by its successful applications to Pb(Zr, Ti)O_3 solid solutions. The lattice parameters and polarization magnitudes of $PbTiO_3$, $PbZrO_3$, and $Pb(Zr_{0.5}Ti_{0.5})O_3$ are well predicted by a single shell model (figure 3.2(d)).

3.2.2 Bond valence model

Based on the bond valence theory (Brown and Shannon 1973, Brown and Wu 1976, Liu *et al* 2013a, Liu *et al* 2013b), the bond valence V_{ij} of a chemical bond between atom i and atom j can be calculated from the bond length r_{ij} with

$$V_{ij} = \left(\frac{r_{0,\,ij}}{r_{ij}}\right)^{C_{ij}} \tag{3.5}$$

where $r_{0,ij}$ and C_{ij} are Brown's empirical parameters that are available for many atomic pairs (Brese and O'Keeffe 1991). The total energy of the system in the bond valence model can be written as the sum of the Coulomb energy (E_{coulomb}), the short-range repulsion energy (E_{r}), the bond-valence energy (E_{BV}), the bond-valence vector energy (E_{BVV}), and the angle potential (E_{a}):

$$E = E_{\mathrm{Coulomb}} + E_{\mathrm{r}} + E_{\mathrm{BV}} + E_{\mathrm{BVV}} + E_{\mathrm{a}} \tag{3.6}$$

$$E_{\mathrm{Coulomb}} = \sum_{i<j} \frac{q_i q_j}{r_{ij}} \tag{3.7}$$

$$E_{\mathrm{r}} = \sum_{i<j} \left(\frac{B_{ij}}{r_{ij}}\right)^{12} \tag{3.8}$$

$$E_{\mathrm{BV}} = \sum_i S_i (V_i - V_{0,i})^2 \tag{3.9}$$

$$E_{\mathrm{BVV}} = \sum_i D_i \left(\mathbf{W}_i^2 - \mathbf{W}_{0,i}^2\right)^2 \tag{3.10}$$

$$E_{\mathrm{a}} = k \sum_i^{N_{\mathrm{oxygen}}} (\theta_i - 180°)^2 \tag{3.11}$$

where q_i is the ionic charge of atom i, B_{ij} is the short-range repulsion parameter; $V_i = \sum_{j \neq i} V_{ij}$ is the bond-valence sum (BVS) and $\mathbf{W}_i = \sum_{j \neq i} \mathbf{V}_{ij}$ is the bond-valence vector sum (BVVS) around atom i as shown in figure 3.3 (Brown 2009, Harvey *et al* 2006); S_i and D_i are scaling parameters in the unit of energy; k is the spring constant and θ is the O–O–O angle along the common axis of two adjacent oxygen octahedra. The BVS energy represents the energy penalty for both overbonded and under-bonded atoms and the (BVVS) energy reflects the tendency of spontaneous symmetry breaking. It has been shown that the BVS energy and the BVVS energy can be related to the second moment and fourth moment of the local density of

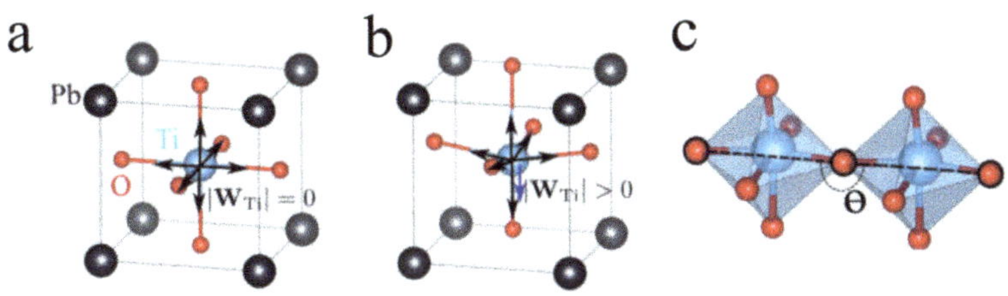

Figure 3.3. Schematic representation of BVVS around Ti in (a) cubic $PbTiO_3$ and (b) tetragonal $PbTiO_3$. Gray, blue, and red spheres denote Pb, Ti, and O. The black arrows scale the individual bond valences, and the blue arrow shows the resultant BVVS W_{Ti}. (c) Angle potential in the bond-valence model. Reprinted from (Liu *et al* 2013c). Copyright © 2013 American Physical Society.

states, respectively, indicating the widely used conservation principles in bond valence theory have formal quantum mechanical justifications.

The bond valence model potential turns out to be quite accurate for many technologically important ferroelectric materials such as $PbTiO_3$ (Liu *et al* 2013b), $BiFeO_3$ (Liu *et al* 2013a), and $BaTiO_3$ (Qi *et al* 2016, Wexler and Rappe 2019), despite using only $\approx$15 force field parameters, as shown in figure 3.4. Taking $PbTiO_3$ as an example, canonical ensemble MD simulations based on the bond valence model predict a ferroelectric-to-paraelectric first-order phase transition at 830 K, comparable to experimental value of 765 K. Meanwhile, the domain wall energy of 180° and 90° walls computed from bond valence MD is 208 mJ m^{-2} and 90 mJ m^{-2}, agreeing well with first-principles results, 170 mJ m^{-2} and 64 mJ m^{-2}, respectively. The bond valence model of $BaTiO_3$ accurately predicts the lattice parameters of rhombohedral, orthorhombic, tetragonal, and cubic phases, with errors lower than 0.5% compared with DFT values. Another highlight of the bond valence model is

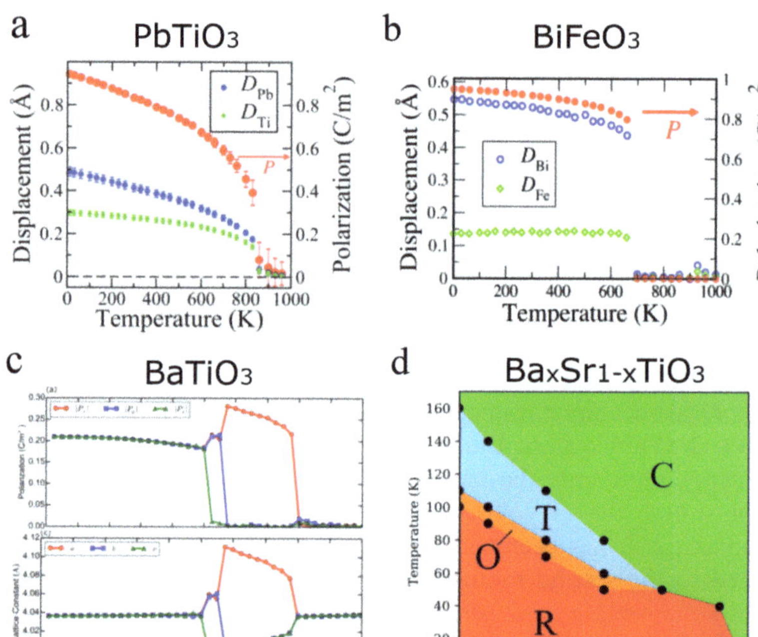

Figure 3.4. (a) Spontaneous polarization and atomic displacements along the *c* axis as functions of temperature in $PbTiO_3$. Reprinted from (Liu *et al* 2013b). Copyright © 2013 American Physical Society. (b) Spontaneous polarization and atomic displacements along the [111] axis as a function of temperature in $BiFeO_3$. Reprinted from (Liu *et al* 2013a). Copyright © 2013 IOP Publishing. (c) Temperature dependence of the polarization and lattice constants in $BaTiO_3$. Reprinted from (Qi *et al* 2016). Copyright © 2016 American Physical Society. (d) Temperature-composition phase diagram of $Ba_xSr_{1-x}TiO_3$ solid solutions. Black circles correspond to the simulated results. Red, orange, blue, and green correspond to the extrapolated regions where the rhombohedral, orthorhombic, tetragonal, and cubic phases are stable, respectively. Reprinted from (Wexler and Rappe 2019). Copyright © 2019 American Physical Society.

the transferability. As demonstrated for $Ba_xSr_{1-x}TiO_3$, a single set of potential parameters can accurately predict the temperature-dependent structural properties of solid solutions of different x. Such transferability is desirable for the construction of robust interatomic potential for multicomponent perovskite solid solutions.

3.2.3 Deep potential

As discussed above, the starting point of force field development is to construct an energy function, followed by a parameterization process by fitting to a database of information including quantum mechanical calculations and/or experimental thermodynamic properties. However, there is no guaranty that a force field model applicable to a specific class of material is robust enough to represent the interatomic potential of a different class. Due to the *p*–*d* hybridization, the transition metal–oxygen bonds in ferroelectrics often possess a mixed ionic-covalent character (Cohen 1992), further complicating the force field development (Liu *et al* 2013a, Liu *et al* 2013b, Phillpot *et al* 2007). Using a more sophisticated energy function with a large number of parameters such as as reactive empirical bond-order (REBO) potential (Brenner *et al* 2009) can improve the transferability at the cost of time-consuming and tedious fitting process. All these difficulties of force field development greatly hinder the applications of MD simulations, particularly to new materials.

The applications of ML methodologies open up new opportunities for force field development. Several ML-based force fields have been developed successfully for various materials systems, including organic molecules (Smith *et al* 2017, Manzhos and Carrington 2008), molecular and condensed water (Bartók *et al* 2013, Morawietz *et al* 2016), metals (Eshet *et al* 2010, Botu and Ramprasad 2015, Zhang *et al* 2020), alloys (Andolina *et al* 2020, Zhang *et al* 2019), semiconductors such as silicon (Babaei *et al* 2019, Behler *et al* 2008, Bartók *et al* 2013) and GeTe (Sossa *et al* 2012), and inorganic halide perovskites (Thomas *et al* 2019). Generally, there are two key ingredients in an ML-based force field: a descriptor that captures the local atomic environment and a nonlinear mapping function that relates the descriptor to the local energy contribution. Early in 2007, Behler and Parrinello achieved accurate predictions of energies and forces for bulk silicon using a neural network (NN), refered to as Behler–Parrinello neural network (BPNN) (Jörg and Parrinello 2007). In BPNN, 'symmetry functions' are used to describe the local geometric environment of an atom and to evaluate the atomic contribution to the total energy. Another successful case is the gradient-domain machine learning (GDML) method (Chmiela *et al* 2017) in which the local atomic environment is captured by eigenvalues of the Coulomb matrix.

Here, the focus is mainly on deep potential molecular dynamics (DPMD) (Zhang *et al* 2018). As illustrated in figures 3.5(a) and (b), in the deep potential (DP) model, the total energy E is defined as the sum of atomic energies E^i parameterized with a deep neural network (DNN), $E^i = E^{\omega_{\alpha_i}}(\mathcal{R}^i)$. Here $\mathcal{R}^i$ is the local environment of atom i in terms of Cartesian coordinates relative to its neighbors within a cutoff radius r_c, α_i denotes the chemical species of ith atom, and ω_{α_i} is the DNN parameter set. Each sub-network of E_i consists of an encoding and a fitting NN with the former

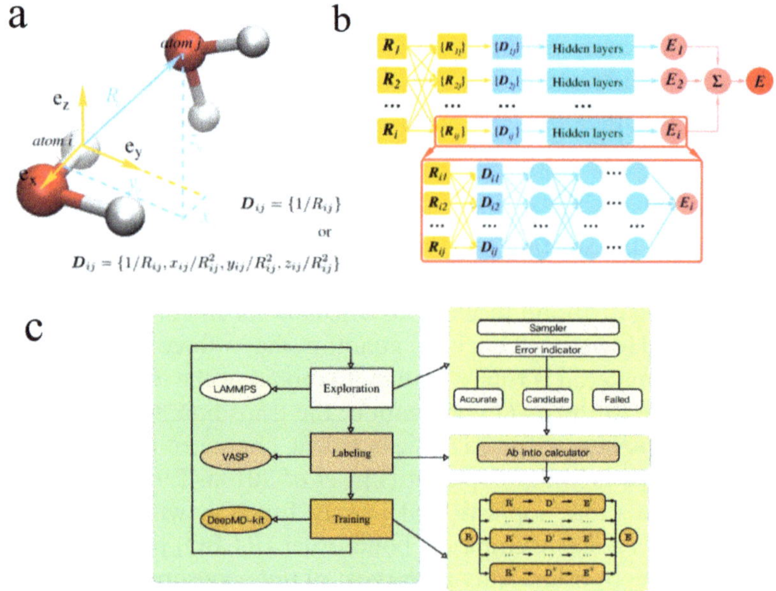

Figure 3.5. (a) Schematic plot of the NN input for the environment of atom *i*, taking water as an example. (b) Schematic plot of the DPMD model. The frame in the box is an enlargement of a DNN. Reprinted from (Zhang *et al* 2018). Copyright © 2018 American Physical Society. (c) DP-GEN workflow. One cycle of DP-GEN contains three steps: exploration, labeling, and training. Molecular dynamics simulations using a DP model are performed in the exploration step to sample new configurations, among which candidate configurations are selected by the error indicator. The labeling step undertakes *ab initio* calculations for the candidate configurations obtained in the exploration step. An ensemble of new DP models are then retrained using the updated training database. Reprinted from (Wu *et al* 2021a). Copyright © 2021 American Physical Society.

maps $\mathcal{R}^i$ to a symmetry-preserved feature matrix $\mathcal{D}^i$ and the latter being a standard feedforward NN. The training process aims to minimize the loss function defined as

$$L(p_\epsilon, p_{\rm f}, p_\xi) = p_\epsilon \Delta\epsilon^2 + \frac{p_{\rm f}}{3N}\sum_i |\, \Delta \boldsymbol{F}_i \,| + \frac{p_\xi}{9}||\Delta\xi||^2 \tag{3.12}$$

where Δ is the difference between the DP prediction and the training data, N is the number of atoms, ϵ is the energy per atom, $\boldsymbol{F}_i$ is the atomic force of atom i, and ξ is the virial tensor divided by N. p_ϵ, p_ξ, and $p_{\rm f}$ are tunable prefactors controlling the weights of energy, force, and stress information used in training, and can be changed during the learning process.

In conventional force field development, the construction of a database for fitting has been rather tricky. It is not so easy to decide what configurations should be included. Given that the training data are usually obtained with expensive first-principles calculations, it is highly desirable to construct an optimal and minimal set of training data. The deep potential generator (DP-GEN) scheme is a concurrent learning procedure that enables efficient updates of first-principles-based training database (figure 3.5(c)). Each closed cycle in DP-GEN consists of three steps: exploration,

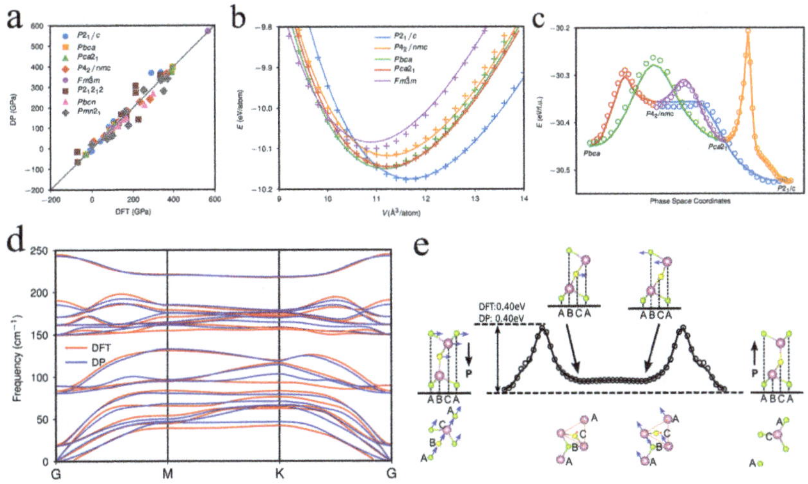

Figure 3.6. (a) Comparison of DFT and DP results on elastic constants of different phases of HfO_2. (b) Equation of states of different HfO_2 polymorphs. Solid lines and cross points denote DFT and DP results, respectively. (c) Comparison of phase transition barriers predicted by DFT (solid line) and DP (empty circle). Reprinted from (Wu *et al* 2021a). Copyright © 2021 American Physical Society. (d) Phonon dispersion relations of monolayer α-In_2Se_3 computed with DFT and DP. (e) DFT and DP energy profiles for polarization reversal pathways in monolayer α In_2Se_3 with multistep switching via concerted motions of upper Se–In–Se layers. Reprinted from (Wu *et al* 2021b). Copyright © 2021 American Physical Society.

labeling and training. Starting with a database, an ensemble of DP models are trained using different initial values of ω_{α_i}. In the following *exploration* step, a DP model is used for MD simulations to explore the configuration space. For each newly sampled configuration $\mathcal{R}_t$ from MD, the ensemble of DP models will generate an ensemble of predictions (e.g. energy and atomic force), the maximum standard deviation of which serves as the criteria for *labeling*: a sampled configuration resulting in large model deviation is labeled for DFT calculations and will be added to the training database for *training* in the next cycle. The cycles will repeat until all sampled configurations from MD are well described by the ensemble of DP models.

We have developed DP models for two newly discovered ferroelectrics, HfO_2 (Wu *et al* 2021a) and α-In_2Se_3 (Wu *et al* 2021b). As shown in figure 3.6, both models have demonstrated accuracy comparable with DFT, capable of predicting a wide range of structural properties such as elastic constants, equation of states, phonon dispersion relationships, and dynamic properties such as phase transition barriers and Curie temperature.

3.3 Phase field model

Before the PhFM method, the sharp interface model was used to simulate the microstructure. The problem with the sharp interface model is that for many physical problems, the features of the phase interface are mostly unknown. Additionally, the phase transition boundary is often not sharp and may posses a complex topology. One of the key ideas of PhFM is to use the gradient of the order parameter to trace the interface, greatly saving the effort in dealing with the interface mesh. The system evolves to minimize the total free energy, which makes the PhFM a powerful technique to study the coupling of multiple fields: regardless of the type of the physical field, its effects can be incorporated by adding the relevant free energy term into the total free energy.

In the description of thermodynamics of phase transition, the thermodynamic function should be chosen with respect to desirable independent variables. For example, the Helmholtz free energy $F(\eta_{ij}, P_i)$ is used to incorporate the strain effects on ferroelectric phase transitions, where the independent variables are strain tensor η_{ij} and polarization P_i. Based on Ginzburg–Landau theory, the free energy is represented continuously in terms of order parameter. These thermodynamic functions are usually polynomials of independent variables for homogeneous phase and can be extended to describe inhomogeneous phase straightforwardly. The spatial distribution of the order parameter is responsible for the inhomogeneous properties in the interface. The free energy could be obtained by adding all the considered energy contributions related to the order parameter together. The time evolution of order parameter is then described by the time-dependent Ginzburg–Landau equations (Chen 2008),

$$\frac{\partial P_i}{\partial t} = -\mathcal{L}\frac{\delta F}{\delta P_i} \quad (i = 1, 3) \tag{3.13}$$

$$F_{\text{total}} = \int_V \left[f_{\text{Landau}}(P_i) + f_{\text{gradient}}(\nabla P_i) + f_{\text{Electric}}(P_i, E_i) + f_{\text{Elastic}}\left(P_i, \eta_{ij}\right)\right] dV \tag{3.14}$$

where $\mathcal{L}$ is the kinetic coefficient related to the mobility of the domain wall and t is the evolution time. The total free energy F_{total} consists of mechanical, electrostatic, Landau chemical, and polarization gradient energies as elaborated in figure 3.7.

Exotic topological polar structures including closure domains, vortices (Yadav 2016), and polar skyrmions (Das 2019) and their associated functionalities in ferroelectric/paraelectric superlattices of $PbTiO_3/SrTiO_3$ have been simulated with PhFM, yielding results comparable with experimental observations. The emergence of various topological polar structures is due a delicate balance between strain, depolarization and gradient energies. Theoretical investigations with PMF prove that the vortex array is the low-energy state for a range of superlattice periods, arsing from the competition between elastic energy due to lattice mismatch, electrostatic energy associated with the built-in depolarization field, and the gradient energy due to polarization rotation as shown in figure 4 in (Yadav 2016). Similarly, by varying the magnitudes of different energy terms, bubble-shaped nano-domains and skyrmion-like topological structures can be obtained as shown in extended data figure 4 in (Das 2019). These topological polar structures may acquire novel, emergent properties. For example, the polar skyrmion bubble phase could give rise to enhanced susceptibilities and negative capacitance (Parkin 2008, Fert 2013). Gaining fundamental knowledge about polar vortices and skyrmions can result in new technological applications such

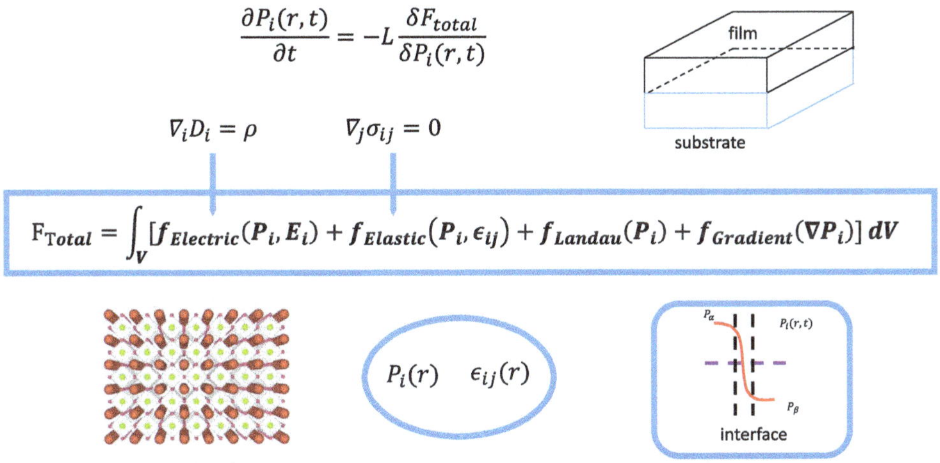

Figure 3.7. A sketch of time-dependent Ginzburg–Landau equations. The total free energy (F_{total}) is described in terms of polarization (P_i) and strain (η_{ij}). f_{Landau}, f_{Gradient}, f_{Electric}, and f_{Elastic} is the stress-free bulk free energy density, gradient energy density, electrostatic energy density, and elastic energy density, respectively.

as 'skyrmionics' in which the data are encoded using the chirality of the individual vortex/skyrmion. It is expected that for novel devices based on polar skyrmions/vortices, it is critical to understand their field-driven dynamical responses so as to switch their chirality via external voltage (Jaafar 2010, Beardsley 2017, Yamada 2007).

3.4 Machine learning

The applications of ML methodologies for computational materials by design have become increasingly popular. This section briefly discusses three representative examples that utilize ML methods for improved design and understanding of ferroelectrics. Compositional variation is an important tool to design and optimize the properties of multicomponent solid solutions. However, the design space increases rapidly with increasing number of components such that it is time-consuming to identify the optimal composition using the conventional 'trial-and-error' approach. Balachandran *et al* developed a two-step ML approach to guide experiments in search of perovskite solid solutions with high ferroelectric Curie temperature (T_C). The system of interest is $x\text{Bi}[\text{Me}'_y\text{Me}''_{(1-y)}]\text{O}_3$–$(1-x)\text{PbTiO}_3$ where Me′ and Me″ are di-, tri-, tetra- or pentavalent cations that occupy the B site of ABO_3 perovksite structure. It is evident that this solid solution exhibits a huge chemical and configuration space: there are 61 500 compositions and only 167 of them are studied experimentally. Moreover, not every arbitrary composition can be synthesized in the perovskite phase. To address this challenge, two independent ML models were developed for classification learning and regression, respectively, using available data in literature. The classification learning models were used to screen compositions giving perovskite structure, the T_C of which was then predicted with regression models. Notably, high-T_C candidates predicted with the two-step ML approach were then synthesized and characterized, the information of which was added to the training set to further improve ML models. The iterative two-step ML approach found a few new compositions that form perovskite structure and exhibit high T_C, as confirmed in experiments. For example, $0.2Bi(Fe_{0.12}Co_{0.88}O_{3-0.8}PbTiO_3$ has a T_C of 898 K, comparable to the state-of-the-art compositions $BiFeO_3$–$PbTiO_3$ and Bi(Zn,Ti)O_3–$PbTiO_3$. In this work, ML methods mapped readily accessible microscopic descriptors such as tolerance factor, valence electron number and Mendeleev number to macroscopic properties (phase stability and T_C) that are either difficult to compute from first-principles or expensive to measure in experiments, so as to facilitate materials screening. This is a typical example of ML-accelerated materials design in which ML methods bridge different scales.

Li *et al* applied principal components analysis (PCA) to data mining multiple realizations of collective dynamics measured through piezoelectric relaxation studies, which enabled rapid determination of the onset temperature of phase transition for a relaxor crystal without the need to measure (or know) the order parameter. They first performed PCA on the 2D data set $AS = AS([x, y, T], t)$ where AS is the amplitude signal from piezoresponse force microscopy (PFM), t is the time dimension, and T is the temperature dimension. The singular value decompositon (SVD) method was employed to obtain eigenvectors and eigenvalues of AS, and the evolution of the eigenvalues with respect to temperature was used to determine phase transition temperature. The results are shown in figure 2 in (Li *et al* 2018). Furthermore, they performed k-means clustering on all the amplitude data sets $AS(V, T, t)$ measured at different voltages (V) and temperatures to map out the temperature-bias phase diagram. This work provides a clear demonstration of the use of unsupervised ML on raw experimental data for automatic determination of phase diagrams without *a priori* knowledge of the underpinning physical mechanisms of phase transitions.

Yadav *et al* (2021) developed a multilevel model for the prediction of T_C, coercive field (E_c), and polarization (P) of $PbTiO_3$-derived ferroelectric solid solutions from compositions. Since P can be obtained directly by first-principle calculations, the work set T_C and E_c as target properties and used several levels of intermediate properties to relate them to solid-solution compositions. They chose electronegativity (χ), ionic radius (R), and valence (Va) as the first, electronic-structure-based and sub-Angstrom, level characteristics; local ionic displacements (D_i) as the second, ion-based, and Angstrom-scale level of properties; P as the third, collective level property. Figure 3.8 shows the simple structure of the multilevel network. Their results proved that to obtain accurate prediction of T_C, A-site and B-site displacements and the valence of the B-site cations are important features.

3.5 Multiscale modeling of ferroelectric switching

Multiscale modeling assumes that physical quantities derived from finer scales could be better than those obtained from phenomenological modeling at a coarser scale. This is because phenomenological modeling is usually empirical while the

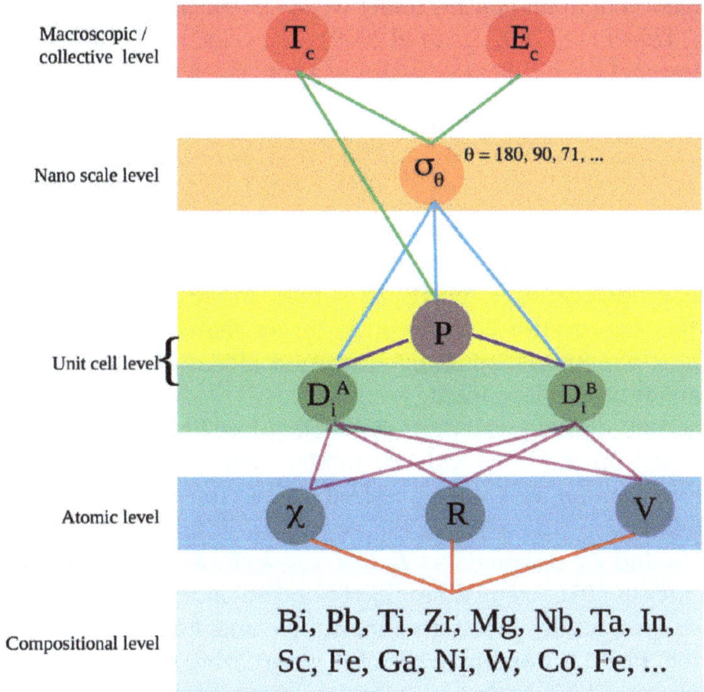

Figure 3.8. Multilevel network relating composition and atomic properties (electronegativity (χ), ionic radius (R), and valence (V)) and macroscopic properties (such as polarization (P), domain-wall energy (σ), ferroelectric-to-paraelectric phase transition temperature (T_C), and coercive field (E_C)). Taken from (Yadav *et al* 2021). Copyright © 2021 Wiley Online Library.

constitutive equations at finer scales are more fundamental but, in the meantime, the relation between the fine and coarse scales is subtle. For simple properties, it is easy to link fine and coarse scales using statistical mechanics. However, the situation is different for heterogeneous systems, particularly for describing large deformation and failure of materials where the history-dependent mechanisms become important. In general, multiscale methods could be divided into two categories, upscaling and resolved-scale methods (Fish *et al* 2021). In upscaling methods, only the average effect is captured because of the approximation at fine scales. The following discussion will focus on upscaling methods, using ferroelectric switching as an example.

Ferroelectric switching is the reversal of polarization in material, which is related to the coercive field extracted from the ferroelectric polarization (P)-electric field (E) hysteresis loop (P–E loop). The polarization reversal in most cases is realized through domain wall motions driven by external electric fields, which shows strong dependence on the strength and frequency of applied fields. It is evident that an accurate prediction of coercive field requires a quantitative understanding of the relationship between domain wall mobility, temperature, and driving electric field. This will help to build a complete understanding of domain-wall-related dynamical properties.

The propagation of domain walls under weak electric fields can be described with a creep process (Tybell *et al* 2002, Jo *et al* 2009). The velocity of domain-wall v depends on the temperature (T) and field (E_e) and is given as

$$v \propto \exp\left[-\frac{U}{k_B T}\left(\frac{E_{C0}}{E_e}\right)^{\mu}\right], \tag{3.15}$$

where U is a characteristic energy barrier, k_B is Boltzmann's constant, E_{C0} is a critical field at which depinning occurs at 0 K, and μ is the dynamical exponent. The presence of defects is implicitly assumed to be responsible for the creep behavior. When the electric field becomes larger than the crossing field E_{C0}, the wall experiences a pinning-depinning transition (Jo *et al* 2009) with the velocity becoming temperature-independent.

In 2007, the 180° domain wall motion in defect-free $PbTiO_3$ was studied with MD simulations based on the bond valence model (Shin *et al* 2007). The MD results revealed that the domain wall motion indeed follows the Merz's law and the velocity can be expressed as $v = v_0 \exp(-E_a/E)$, where v_0 is the coefficient with $\mu = 1$ in equation (3.15) and E_a is defined as $E_a = UE_{C0}/k_B T$, as shown in figure 2 in (Shin *et al* 2007). Later in 2016, using a modified bond valence model that includes BVVS energy, Liu *et al* carefully studied temperature- and field-dependent domain-wall velocity for 90° $PbTiO_3$ domain walls (Liu *et al* 2016b), revealing two regions as shown in figure 3.9. The domain wall exhibits a temperature-sensitive creep behavior

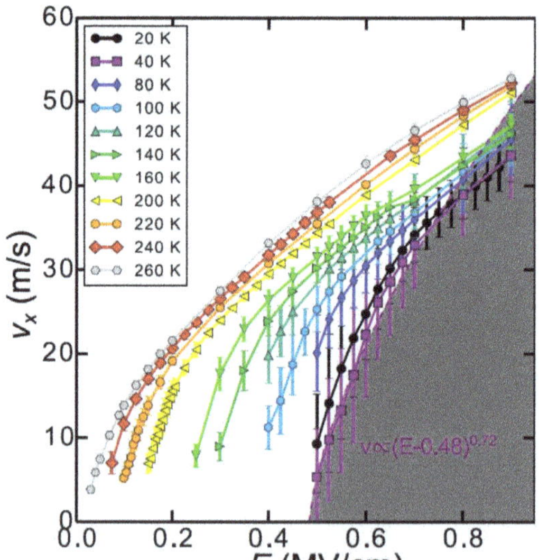

Figure 3.9. Temperature and field-dependent domain-wall speeds from molecular dynamics simulations. The 90° domain-wall mobility as a function of electric field under different temperatures. Temperature and field-dependent data suggest an intrinsic creep-depinning transition. The domain-wall velocity data at 40 K are in the flow region (shaded area) giving $\theta = 0.72$ and $E_{C0} = 0.482$ MV m^{-1}. Reprinted from (Liu *et al* 2016b). Copyright © 2016 Springer Nature.

following the Merz's law under low-field conditions. In the high-field region, the influence of temperature becomes weak, indicating a depinning character. The results also suggest that the presence of defects is not necessary for the intrinsic creep-depinning transition. Instead, the nucleation-and-growth mechanism is responsible for such transition. At low fields, the nucleation with a field-dependent barrier is the rate determining step of domain wall motion, and the thermal activation is important for the nucleation. At high fields, the nucleation barrier approaches zero such that the domain wall motion becomes temperature insensitive and its velocity mainly depends on the field strength.

Another important insight gained from MD simulations concerns the shape of the nucleus. In the seminal work of Miller and Weinreich, a nucleation-and-growth mechanism was proposed to explain the electric field-dependence of domain-wall velocity $v(E)$ (Miller and Weinreich 1960). They assumed the nucleus boundary is sharp so that the depolarization-contributed domain wall energy is much larger than the interface energy. A more detailed atomistic picture was obtained from the MD simulations of ferroelastic 90° domain walls in $PbTiO_3$ (Liu *et al* 2016b), revealing a diamond-like nucleus with substantial diffuseness at the boundary characterized by a gradual polarization change, as shown in figure 3.10. In contrast to the Miller–Weinreich theory, the polarization changes gradually at the diamond boundary,

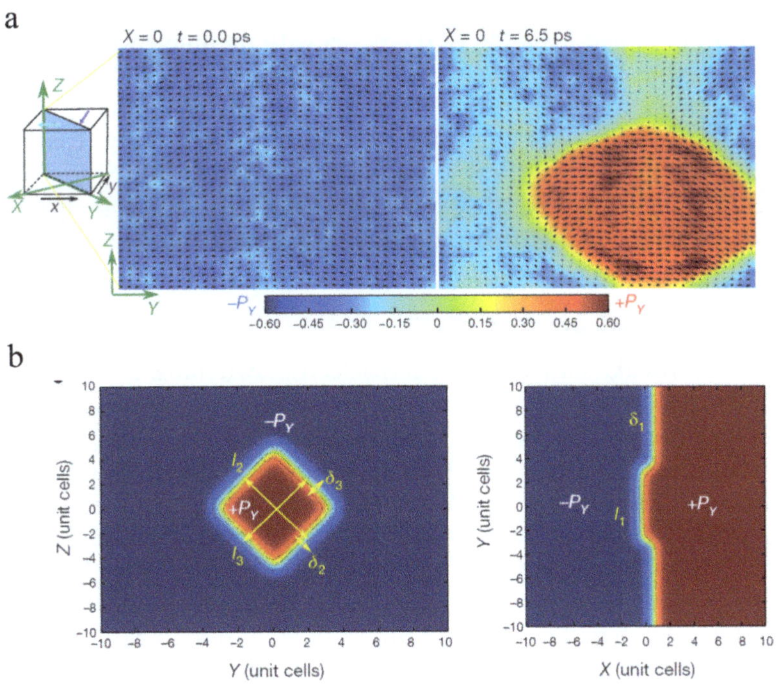

Figure 3.10. (a) Nucleation process at a 90° domain wall. In the presence of electric field, a diamond-like nucleus forms at t = 6.5 ps. Adapted from (Liu *et al* 2016b). Copyright © 2016 Springer Nature. (b) Polarization profile of a nucleus generated by analytical nucleation model.

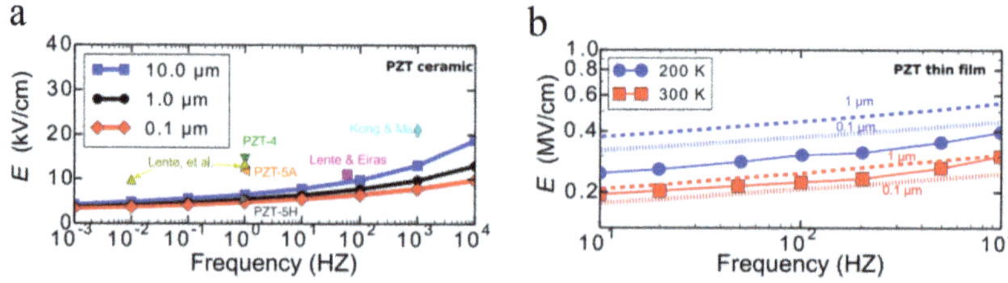

Figure 3.11. (a) Simulated frequency dependence of coercive fields E_c for PZT ceramics for various domain sizes (see legend) at 300 K. (b) Simulated frequency- and temperature-dependent coercive fields for PZT thin films. Adapted from (Liu *et al* 2016b). Copyright © 2018 Springer Nature.

leading to a much smaller depolarization charge. The local interface energy plays the dominant role and governs the energetics of nucleation and growth.

Based on the Landau–Ginzburg–Devonshire (LGD) theory, a universal, nucleation-and-growth-based analytical model that incorporates MD insights was developed to quantify the nucleation kinetics for many types of domain walls. With this analytical nucleation model, one can easily map out the energy of the nucleus as a function of nucleus size and estimate the nucleation activation energy (ΔU_{nuc}). The Avrami theory of transformation kinetics was used to relate ΔU_{nuc} to the activation field E_a in Merz's law. For a given domain wall, all input parameters for the analytical model can be directly estimated from DFT calculation. Liu *et al* studied polarization reversal of $PbTiO_3$ with this analytical model (Liu *et al* 2016b). The value of E_a obtained from the LGD model with parameters (Zhao and Truhlar 2008) estimated with DFT PBEsol calculations (Perdew *et al* 2008). The theoretical coercive fields as shown in figure 3.11 using the parameters of the 90° domain wall agree well with experimental E_c values (5–20 kV cm^{-1}) over a large frequency range (Rino *et al* 2004, Lente and Eiras 2001, Kong and Ma 2001, Shrout and Zhang 2007). The E_c values based on 180° the domain wall are larger, in agreement with thin-film results (Yang *et al* 2010).

3.6 Challenges and perspectives

The case of multiscale simulation of ferroelectric switching is a typical example where careful MD studies of representative ferroelectric materials were performed to extract the fundamental factors that govern macroscopic materials properties. The shape of the nucleus revealed from MD simulations is critical for the development of a physically justified phase field model for accurate E_c predictions. Within the upscaling multiscale approach, MD plays a pivotal role. Due to the difficulty of force field development, the current multiscale approach remains largely 'case-by-case', only applicable to a single material. Recent applications of ML methodologies to force field development are changing this state of affairs. In particular, for the DP method, thanks to the ability of deep neural work to faithfully represent complex and highly nonlinear potential energy surface, it is now plausible to develop a transferable and expandable force field library. We suggest that the training database associated with the DP model is a more fundamental entity. Therefore, by maintaining and sharing the database, force fields applicable to new materials systems can be further developed based on available ones by adding new configurations to the existing database. With ever increasing computing power, we believe ML-assisted high-throughput force field development is within reach.

Alternative to physics-based upscaling multiscale simulation, ML-assisted computational materials by design discussed in section 3.4 can also be considered as a way to bridge different scales and to predict macroscopic properties that are relevant for applications. Such a data-driven method is ideal to faithfully link microscopic variables at fine scales to governing equations at coarser scales, if provided enough accurate data. We thus propose a hybrid multiscale simulation that combines the advantages of physics-based and data-driven methods. The physics-based methods such as DFT, MD, and PhFM can generate a large amount of data spanning different time and length scales to complement limited experimental data. The hybrid database can then be used to train ML models that map easily accessible descriptors to macroscopic properties of interest. It is expected that new ML algorithms combined with increasing availability of large computational resources may help to reveal relationships between data sets unobtainable from physical insight alone (Butler *et al* 2018). We believe the hybrid multiscale simulation is a promising approach to facilitate the discovery and design of complex multicomponent ferroelectrics.

Acknowledgments

The authors acknowledge support from National Natural Science Foundation of China (12074319) and Westlake Education Foundation.

References

Alder B J and Wainwright T E 1959 Studies in molecular dynamics. I. General method *J. Chem. Phys.* **31** 459

Anderson P W 1972 More is different *Science* **177** 393

Andolina C M, Williamson P and Saidi W A 2020 Optimization and validation of a deep learning CuZr atomistic potential: robust applications for crystalline and amorphous phases with near-DFT accuracy *J. Chem. Phys.* **152** 154701

Babaei H, Guo R, Hashemi A and Lee S 2019 Machine-learning-based interatomic potential for phonon transport in perfect crystalline Si and crystalline Si with vacancies *Phys. Rev. Mater* **3** 074603

Bartók A P, Gillan M J, Manby F R and Gábor C 2013 Machine-learning approach for one- and two-body corrections to density functional theory: applications to molecular and condensed water *Phys. Rev.* B **88** 054104

Bartók A P, Kermode J, Bernstein N and Csányi G 2018 Machine learning a general-purpose interatomic potential for silicon *Phys. Rev.* X **8** 041048

Behler J, Martonák R, Donadio D and Parrinello M 2008 Pressure-induced phase transitions in silicon studied by neural network-based metadynamics simulations *Phys. Status Solidi* B **245** 2618

Beardsley R P 2017 Deterministic control of magnetic vortex wall chirality by electric field *Sci. Rep.* **7** 7613

Bellaiche L, Garcia A and Vanderbilt D 2000 Finite-temperature properties of $Pb(Zr_{1-x}Ti_x)O_3$ alloys from first principles *Phys. Rev. Lett.* **84** 5427

Botu V and Ramprasad R 2015 Learning scheme to predict atomic forces and accelerate materials simulations *Phys. Rev.* B **92** 094306

Brenner D W, Shenderova O A, Harrison J A, Stuart S J, Ni B and Sinnott S B 2002 A second-generation reactive empirical bond order (REBO) potential energy expression for hydrocarbons *J. Phys.: Condens. Matter* **14** 783

Brese N E and O'Keeffe M 1991 Bond-valence parameters for solids *Acta Crystallogr.* B **47** 192

Brown I D 2009 Recent developments in the methods and applications of the bond valence model *Chem. Rev.* **109** 6858

Brown I D and Shannon R D 1973 Empirical bond-strength–bond-length curves for oxides *Acta Crystallogr.* A **29** 266

Brown I D and Wu K K 1976 Empirical parameters for calculating cation–oxygen bond valences *Acta Crystallogr.* B **32** 1957

Butler K T, Davies D W, Cartwright H, Isayev O and Walsh A 2018 Machine learning for molecular and materials science *Nature* **559** 547

Chen L-Q 2008 Phase-field method of phase transitions/domain structures in ferroelectric thin films: a review *J. Am. Ceram. Soc.* **91** 1835

Chmiela S, Tkatchenko A, Sauceda H E, Poltavsky I, Schütt K T and Müller K-R 2017 Machine learning of accurate energy-conserving molecular force fields *Sci. Adv.* **3** 1603015

Cohen R E 1992 Origin of ferroelectricity in perovskite oxides *Nature* **358** 136

Das S 2019 Observation of room-temperature polar skyrmions *Nature* **568** 368

de Gironcoli S, Baroni S and Resta R 1989 Piezoelectric properties of III-V semiconductors from first-principles linear-response theory *Phys. Rev. Lett.* **62** 2853

de Jong M, Chen W, Geerlings H, Asta M and Persson K A 2015 A database to enable discovery and design of piezoelectric materials *Sci. Data* **2** 150053

Eshet H, Khaliullin R Z, Kühne T D, Behler J and Parrinello M 2010 *Ab initio* quality neural-network potential for sodium *Phys. Rev.* B **81** 184107

Fert A 2013 Skyrmions on the track *Nat. Nanotechnol* **8** 152

Fish J, Wagner G J and Keten S 2021 Mesoscopic and multiscale modelling in materials *Nat. Mater.* **20** 774

Garrity K F, Bennett J W, Rabe K M and Vanderbilt D 2014 Pseudopotentials for high-throughput DFT calculations *Comput. Mater. Sci.* **81** 446

Gindele O, Kimmel A, Cain M G and Duffy D 2015 Shell model force field for lead zirconate titanate $Pb(Zr_{1-x}Ti_x)O_3$ *J. Phys. Chem.* C **119** 17784

Graf M, Sepliarsky M, Tinte S and Stachiotti M G 2014 Phase transitions and antiferroelectricity in $BiFeO_3$ from atomic-level simulations *Phys. Rev.* B **90** 184108

Harvey M A, Baggio S and Baggio R 2006 A new simplifying approach to molecular geometry description: the vectorial bond-valence model *Acta Crystallogr.* B **62** 1038

Jaafar M 2010 Control of the chirality and polarity of magnetic vortices in triangular nanodots *Phys. Rev.* B **81** 054439

Jain A, Ong S P and Hautier G *et al* 2013 The materials project: a materials genome approach to accelerating materials innovation *APL Mater* **1** 011002

Jo J Y, Yang S M, Kim T H, Lee H N, Yoon J-G, Park S, Jo Y, Jung M H and Noh T W 2009 Nonlinear dynamics of domain-wall propagation in epitaxial ferroelectric thin film *Phys. Rev. Lett.* **102** 045701

Jörg B and Parrinello M 2007 Generalized neural-network representation of high-dimensional potential-energy surfaces *Phys. Rev. Lett.* **98** 146401

Kong L B and Ma J 2001 PZT ceramics formed directly from oxides via reactive sintering *Mater. Lett.* **51** 95

Lebeugle D, Colson D, Forget A and Viret M 2007 Very large spontaneous electric polarization in $BiFeO_3$ single crystals at room temperature and its evolution under cycling fields *Appl. Phys. Lett.* **91** 022907

Lente M H and Eiras J A 2001 90° Domain reorientation and domain wall rearrangement in lead zirconate titanate ceramics characterized by transient current and hysteresis loop measurements *J. Appl. Phys.* **89** 5093

Li L, Yang Y, Zhang D, Ye Z-G, Jesse S, Kalinin S V and Vasudevan R K 2018 Machine learning-enabled identification of material phase transitions based on experimental data: exploring collective dynamics in ferroelectric relaxors *Sci. Adv.* **4** 8672

Liu S, Grinberg I and Rappe A M 2013a Development of a bond-valence based interatomic potential for $BiFeO_3$ for accurate molecular dynamics simulations *J. Physics.: Condens. Matter* **25** 102202

Liu S, Grinberg I and Rappe A M 2016b Intrinsic ferroelectric switching from first principles *Nature* **534** 360

Liu S, Grinberg I, Takenaka H and Rappe A M 2013b Reinterpretation of the bond-valence model with bond-order formalism: An improved bond-valence-based interatomic potential for $PbTiO_3$ *Phys. Rev.* B **88** 104102

Liu S, Grinberg I, Takenaka H and Rappe A M 2013c Reinterpretation of the bond-valence model with bond-order formalism: An improved bond-valence-based interatomic potential for $PbTiO_3$ *Phys. Rev.* B **88** 104102

Liu J, Fernández-Serra M V and Philip B A 2016a First-principles study of pyroelectricity in GaN and ZnO *Phys. Rev.* B **93** 081205

Manzhos S and Carrington T 2008 Using neural networks, optimized coordinates, and high-dimensional model representations to obtain a vinyl bromide potential surface *J. Chem. Phys.* **129** 224104

Metropolis N, Rosenbluth A W, Rosenbluth M N and Teller A H 1953 Equation of state calculations by fast computing machines *J. Chem. Phys.* **21** 1087

Miller R C and Weinreich G 1960 Mechanism for the sidewise motion of 180° domain walls in barium titanate *Phys. Rev.* **117** 1460

Mitchell P J and Fincham D 1993 Shell model simulations by adiabatic dynamics *J. Phys.: Condens. Matter* **5** 1031

Morawietz T, Singraber A, Dellago C and Behler J 2016 How van der Waals interactions determine the unique properties of water *Proc. Natl Acad. Sci.* **113** 8368

Parkin S S P 2008 Magnetic domain-wall racetrack memory *Science* **320** 190

Perdew J P, Ruzsinszky A, Csonka G I, Vydrov O A, Scuseria G E, Constantin L A, Zhou X and Burke K 2008 Restoring the density-gradient expansion for exchange in solids and surfaces *Phys. Rev. Lett.* **100** 136406

Phillpot S R, Sinnott S B and Asthagiri A 2007 Atomic-level simulation of ferroelectricity in oxides: current status and opportunities *Annu. Rev. Mater. Res.* **37** 239

Qi Y, Liu S, Grinberg I and Rappe A M 2016 Atomistic description for temperature-driven phase transitions in $BaTiO_3$ *Phys. Rev.* B **94** 134308

Rahman A 1964 Correlations in the motion of atoms in liquid argon *Phys. Rev.* **136** A405

Rino J P, Lente M H, Picinin A and Eiras J A 2004 90° Domain wall relaxation and frequency dependence of the coercive field in the ferroelectric switching process *J. Appl. Phys.* **95** 2646
Sághi-Szabó G, Cohen R E and Krakauer H 1998 First-principles study of piezoelectricity in $PbTiO_3$ *Phys. Rev. Lett.* **80** 4321
Sepliarsky M, Asthagiri A, Phillpot S R, Stachiotti M G and Migoni R L 2005 Atomic-level simulation of ferroelectricity in oxide materials *Curr. Opin. Solid State Mater. Sci.* **9** 107
Sepliarsky M and Cohen R E 2002 Development of shell model potential for molecular dynamics for $PbTiO_3$ by fitting first principles results *AIP Conf. Proc.* 626 36
Sepliarsky M and Cohen R E 2011 First-principles based atomistic modeling of phase stability in PMN-PT *J. Phys.: Condens. Matter* **23** 435902
Sepliarsky M, Stachiotti M G and Migoni R L 1995 Structural instabilities in $KTaO_3$ and $KNbO_3$ described by the nonlinear oxygen polarizability model *Phys. Rev.* B **52** 4044
Sepilarsky M, Phillpot S R, Wolf D, Stachiotti M G and Migoni R L 2000 Atomic-level simulation of ferroelectricity in perovskite solid solutions *Appl. Phys. Lett.* **76** 3986
Shimada T, Wakahara K, Ulmeno Y and Kitamura T 2008 Shell model potential for $PbTiO_3$ and its applicability to surfaces and domain walls *J. Phys. Condens. Matter* **20** 325225
Shin Y-H, Grinberg I, Chen I-W and Rappe A M 2007 Nucleation and growth mechanism of ferroelectric domain-wall motion *Nature* **449** 881
Shrout T R and Zhang S J 2007 Lead-free piezoelectric ceramics: alternatives for PZT? *J. Electroceram.* **19** 113
Smit B and Frenkel D 2001 *Understanding Molecular Simulation: From Algorithms to Applications* vol 1 (New York: Academic)
Smith J S, Isayev O and Roitberg A E 2017 ANI-1: an extensible neural network potential with DFT accuracy at force field computational cost *Chem. Sci.* **8** 3192
Sosso G C, Miceli G, Caravati S, Behler J and Bernasconi M 2012 Neural network interatomic potential for the phase change material GeTe *Phys. Rev.* B **85** 174103
Stefano Curtarolo W, Setyawan G L W and Hart M *et al* 2012 AFLOW: an automatic framework for high-throughput materials discovery *Comput. Mater. Sci.* **58** 218
Tan H, Takenaka H, Xu C, Duan W, Grinberg I, Andrew and Rappe M 2018 First-principles studies of the local structure and relaxor behavior of $Pb(Mg_{1/3}\ Nb_{2/3})O_3$-$PbTiO_3$-derived ferroelectric perovskite solid solutions *Phys. Rev.* B **97** 174101
Thomas J C, Bechtel J S, Natarajan A R and Van der Ven A 2019 Machine learning the density functional theory potential energy surface for the inorganic halide perovskite $CsPbBr_3$ *Phys. Rev.* B **100** 134101
Tinte S, Stachiotti M G, Sepliarsky M, Migoni R L and Rodriguez C O 1999 Atomistic modelling of $BaTiO_3$ based on first-principles calculations *J. Phys.: Condens. Matter* **11** 9679
Tinte S, Stachiotti M G, Phillpot S R, Sepliarsky M, Wolf D and Migoni R L 2004 Ferroelectric properties of $Ba_xSr_{1-x}TiO_3$ solid solutions obtained by molecular dynamics simulation *J. Phys.: Condens. Matter* **16** 3495
Tybell T, Paruch P, Giamarchi T and Triscone J-M 2002 Domain wall creep in epitaxial ferroelectric $Pb(Zr_{0.2}\ Ti_{0.8})O_3$ thin films *Phys. Rev. Lett.* **89** 097601
Vielma J M and Schneider G 2013 Shell model of $BaTiO_3$ derived from *ab-initio* total energy calculations *J. Appl. Phys.* **114** 174108
Völker B, Marton P, Elsässer C and Kamlah M 2011 Multiscale modeling for ferroelectric materials: a transition from the atomic level to phase-field modeling *Contin. Mech. Thermodyn* **23** 435

Wexler R B, Qi Y and Rappe A M 2019 Sr-induced dipole scatter in $Ba_xSr_{1-x}TiO_3$: insights from a transferable-bond valence-based interatomic potential *Phys. Rev.* **B 100** 174109

Wu J, Zhang Y, Zhang L and Liu S 2021a Deep learning of accurate force field of ferroelectric HfO_2 *Phys. Rev.* B **103** 024108

Wu J, Bai L, Huang J, Ma L, Liu J and Liu S 2021b Accurate force field of two-dimensional ferroelectrics from deep learning *Phys. Rev.* B **104** 174107

Yadav A K 2016 Observation of polar vortices in oxide superlattices *Nature* **530** 198

Yadav S, Samanta A, Shafir O and Grinberg I 2021 A multi-level analytical theory for prediction of ferroelectric perovskite oxide properties from composition *Adv. Mater.* **34** 2106105

Yamada K 2007 Electrical switching of the vortex core in a magnetic disk *Nat. Mater.* **6** 270

Yang S M, Jo J Y, Kim T H, Yoon J-G, Song T K, Lee H N, Marton Z, Park S, Jo Y and Noh T W 2010 AC dynamics of ferroelectric domains from an investigation of the frequency dependence of hysteresis loops *Phys. Rev.* B **82** 174125

Zhang L, Han J, Wang H, Car R and Weinan E 2018 Deep potential molecular dynamics: a scalable model with the accuracy of quantum mechanics *Phys. Rev. Lett.* **120** 143001

Zhang L, Lin D-Y, Wang H, Car R and Weinan E 2019 Active learning of uniformly accurate interatomic potentials for materials simulation *Phys. Rev.* M **3** 023804

Zhang Y, Wang H, Chen W, Zeng J, Zhang L, Wang H and Weinan E 2020 DP-GEN: a concurrent learning platform for the generation of reliable deep learning based potential energy models *Comput. Phys. Commun.* **253** 107206

Zhao Y and Truhlar D G 2008 Construction of a generalized gradient approximation by restoring the density-gradient expansion and enforcing a tight Lieb–Oxford bound *J. Chem. Phys.* **128** 184109

Chapter 4

Functional simulation of ferroelectric materials

Benjamin Ducharne

Chapters 2 and 3 have discussed microscopic modelling. Nevertheless, ferroelectric materials are eventually applied macroscopically as either bulk materials or films. The dynamic features of the microscale domains and domain walls add further complexity to the dynamic behaviours at the macroscale. Grain boundaries in widely used ferroelectric ceramics can cause even more significant complexity. Consequently, practical ferroelectric materials suffer to an increasing extent from nonlinearity, such as saturation, hysteresis and frequency dependence, which are directly related to the high-amplitude/-frequency driving forces. Development of ferroelectric devices exhibiting such nonlinearity requires precise simulation tools. Macroscopic functional modelling of ferroelectric ceramics has been a research challenge where different approaches are tested. This chapter describes selected solutions to the simulation challenges. Influence of electric field, frequency, temperature, light and mechanical stress on the quasi-static behaviour is discussed. The chapter provides insights to efficient simulation methods within a well-defined experimental frame.

4.1 Nonlinearity in ferroelectrics

During World War II, independent research teams from the USA, USSR and Japan discovered that by poling ferroelectrics, i.e. re-orienting spontaneous polarizations in domains with an external DC electric field to obtain a net polarization, exceptional piezoelectric coupling can be exhibited (Xu 2013). The piezoelectric effect is defined as the change of charge displacement on material surfaces under external mechanical stress or the material deformation caused by applying external electric field (Martin 1972). Since then, the ever popular ferroelectrics $BaTiO_3$ and $Pb(Zr,Ti)O_3$ (PZT) have been developed and the resultant strong piezoelectricity has become one of the major applications for ferroelectrics.

Ferroelectric-based piezoelectric devices such as actuators behave linearly under low-amplitude excitation. This means that by applying electric field within certain range, the material/device deformation has a linear relationship with the external

doi:10.1088/978-0-7503-3975-9ch4

field. However, the linearity will soon become invalid under higher stimuli due to the hysteretic switchability of ferroelectric domains. This is detrimental to the precision of piezoelectric actuators.

The nonlinear behaviour of ferroelectric materials is classically observed in experiments by measuring the response of polarization (*P*) when subject to an electric field (*E*) that varies over time. Such a measurement induces the ferroelectric hysteresis loop (*P–E* loop) (Picinin *et al* 2004). In practice, measuring a cycle of *P–E* loop involves the application of a high-amplitude excitation whilst monitoring the simultaneously resultant current density. *P* is then calculated by integration afterwards.

A *P–E* loop reveals many essential properties and forms the basis for the fundamental understanding of the ferroelectric material in question as well as for potentially critical properties not only in piezoelectric actuators but also ferroelectric memories, piezoelectric transducers, sensors, and energy harvesters (Bowen *et al* 2014, Picinin *et al* 2004, Masui *et al* 2003, Hegewald *et al* 2008).

A hysteresis cycle is dependent on temperature, stress, and frequency. For instance, an instantaneous increase of the compression on the material leads to a reduction of polarization, resulting in the degradation of piezoelectric performance. This degradation is usually considered due to the irreversible motion of domain walls (Lynch 1996, Krueger 1968, Weaver *et al* 2011, Wang *et al* 1998). On the other hand, significant variations of temperature or stress can trigger phase transitions, leading to drastic changes of material properties (Wang *et al* 2011).

The resultant nonlinear and hysteretic features induce a power limitation for heavy-duty transducers or a lack of controllability for sensors and actuators (Hegewald *et al* 2008). Consequently, the nonlinear modelling, including nonlinear considerations of hysteresis and saturation, appears to be a vital issue for understanding/designing device behaviours and thus for determining the boundary conditions in applications (Damjanovic 2006).

Several models have been proposed in the literature for a global understanding of, or to take into account, the hysteresis and nonlinear behaviour of ferroelectric materials (Smith *et al* 2003, Kamlah 2001, Ducharne *et al* 2007, Sutor *et al* 2010, Klinkel 2006, Arlt 1996, Hughes and Wen 1997). Predictive simulation methods such as the phase-field modelling based on thermodynamic relations can be used to predict the distribution of ferroelectric domains in a perfect single crystal specimen (Li *et al* 2001) but these methods are complex, time consuming and non-adaptive to ferroelectric hysteresis of ceramics.

In this chapter, phenomenological models will be described in a limited predictive range but they are able to provide physical insights which help to understand ferroelectric ceramics. Classic experimental conditions for the acquisition of ferroelectric hysteresis cycles are defined first, followed by discussions of selected phenomenological simulation methods.

4.2 Experimental setup for the hysteresis loop measurement

The shape of *P–E* loop depends on the external conditions (temperature, stress, light, etc) as well as on the waveform of the electric field (frequency, amplitude).

The electric charge displacements are not only induced by the external electric field but also by a mechanical stress through the piezoelectric effect or by a temperature change through the pyroelectric effect. Reciprocally, the mechanical strain in a piezoelectric material may be induced by an electric field through the converse piezoelectric effect, by an external stress, and by a temperature change due to the thermal expansion of the material (Lynch 1996, Wang *et al* 1998). These coupling effects imply important experimental constraints on the property measurements. Precise control of all these conditions is required to obtain reproducible results.

In a basic ferroelectric evaluation system as illustrated in figure 4.1, the specimen under test is subjected to an electric field at a constant temperature with free mechanical boundary conditions. The specimen is subjected to a maximum voltage effectively higher than the corresponding coercive field of the material. A high-voltage amplifier driven by a frequency generator is used to precisely control the frequency and amplitude of the input voltage thus electric field (Wang *et al* 2010).

To avoid dielectric breakdown, the specimen under test should be electrically insulated with silicone grease or oil. The electric charge displacement is calculated from the results of the charge measurement and P is computed by integrating the current values. The field-induced strain in the thickness direction (perpendicular to the planes of the specimen's electrodes) can be measured simultaneously using a beam laser interferometer or a capacitive sensor.

When the influence of stress is considered, piezo-stack actuators or traction/compression benches can be used to control the mechanical force (Guyomar *et al* 2008, 2010a). Similarly, at relatively low temperatures, the silicon oil bath temperature can be set precisely to help to evaluate the temperature influence. Instead, a thermal chamber can be used for elevated temperatures. In addition, if the influence of incident light is considered (Ducharne *et al* 2020), a light-emitting diode bulb can be used as the light source. Selection and control of the illumination spectrum should be carried out according to the optical response or band gap of the specimen to be tested.

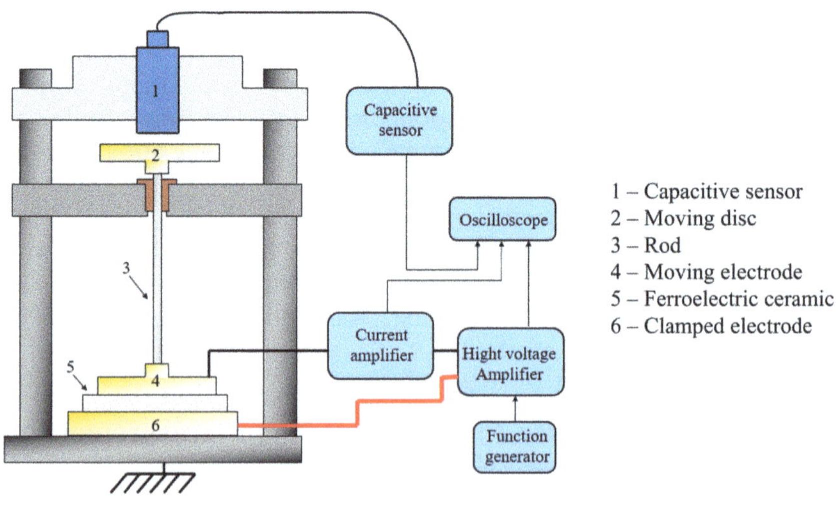

Figure 4.1. Schematic of a typical experimental setup for *P–E* loop measurement.

For relatively low frequencies, an ideal specimen should have a large surface area and a small thickness. Prior to the measurements, the specimen should be annealed at a high temperature to remove possible mechanical stress introduced during sintering and machining. Electrodes are deposited, e.g. via printing or sputtering, on the surfaces of the specimen. For higher frequencies, the surface area can be reduced accordingly, in order to decrease the polarization current and thus to keep the values within the range of the voltage amplifier.

4.3 Simulation methods

Many phenomenological ferroelectric hysteresis simulation methods have already been described in scientific literature (Smith *et al* 2003, Kamlah 2001, Ducharne *et al* 2007, Sutor *et al* 2010, Klinkel 2006, Arlt 1996, Hughes and Wen 1997). This chapter does not intend to present all the methods extensively. Instead, only the most relevant techniques are considered. The behaviour in the low-frequency range is discussed first, followed by the consideration of the frequency effect. The effects of stress, temperature and light will be briefly introduced at the end. It should be noted that the scope of the phenomenological methods in this chapter is limited to where phase transition is absent.

4.3.1 Quasi-static contribution

Hysteresis in a ferroelectric ceramic can be considered as a nonlinear delay separating the excitation electrical field and the tested specimen's polarization answer. Towards the low-frequency range, the polarization variation of a ferroelectric material becomes insensitive to the excitation frequency. This observation gives rise to a cut-off frequency known as the quasi-static threshold. The ferroelectric state beyond this threshold (in the decreasing direction) is called the quasi-static state. Similar phenomena can also be observed in ferromagnetic materials in terms of the ferromagnetic hysteresis.

The Preisach's model

In a quasi-static state (frequency, $f \ll 1$ Hz for a typical soft PZT composition), the cumulative periodic value of the dielectric losses is independent of frequency. Different approaches are available in literature to simulate the quasi-static hysteresis behaviour. Among those approaches, the Preisach's model exhibits particularly exciting properties, including:

- High accuracy in the reconstitution of unsymmetrical phases (first polarization curve, first-order reversal curves, etc);
- A parameter estimation based on a set of experimental information which can be limited to a major hysteresis cycle and a first polarization curve (Biorci and Pescetti 1958);
- Being easily reversible where E can be replaced by P as the model's input with a little extra effort.

In history, Preisach's model has firstly been used to describe the hysteresis phenomenon in ferromagnetic materials (Preisach 1935, Mayergoyz 1986).

More recently, the model has also been validated for ferroelectric materials (Sutor *et al* 2010, Hughes and Wen 1997, Hu and Mrad 2003, Bernard *et al* 2007). Preisach's model assumes that the material polarization is determined by the contribution of a set of elementary hysteresis loops having a distribution function over the Preisach's triangle. More precisely (Mayergoyz 1986), it assumes that the tested material behaviour can be described by a set of elementary hysteresis loops $E(\alpha,\beta)$ where α and β are, respectively, the upper and lower switching fields with $\alpha \geqslant \beta$. The polarization (P) is determined by the contribution of the set of elementary particles having a distribution function of $\psi(\alpha,\beta)$ over the Preisach's triangle in the (α,β)-plane, as explained in figure 4.2 and by the equations below.

$$P = P_{\text{sat}} \cdot \int_{\alpha}\int_{\beta} E(\alpha, \beta) \cdot \psi(\alpha, \beta) \cdot \mathrm{d}\alpha \cdot \mathrm{d}\beta$$

with

$$\int_{\alpha}\int_{\beta} \psi(\alpha, \beta) \cdot \mathrm{d}\alpha \cdot \mathrm{d}\beta = 1$$

and

$$-E_{\text{sat}} \leqslant \alpha \leqslant E_{\text{sat}} \ (\beta \leqslant \alpha)$$

P_{sat} is the saturation polarization value when $E = E_{\text{sat}}$. According to the conception of the classical Preisach's model, the centred cycles of magnitude E_1, E_2 and E_3 shown in figure 4.3 are obtained by integration of the distribution function over, respectively, the triangles 1, 2 and 3 shown in figure 4.2.

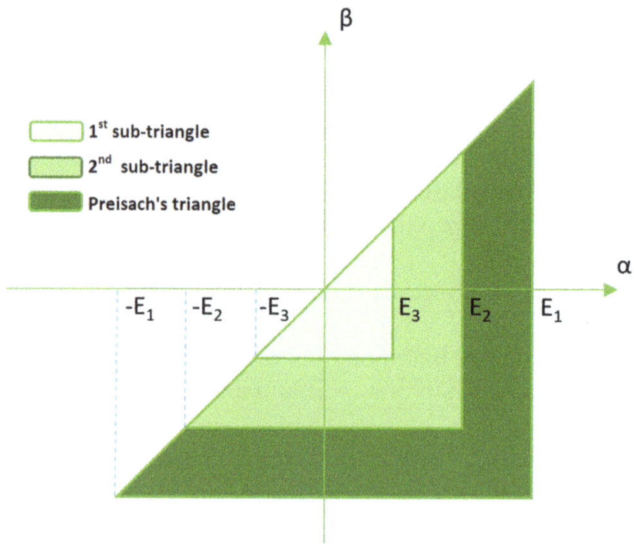

Figure 4.2. The Preisach's triangle with the first and second sub-triangles.

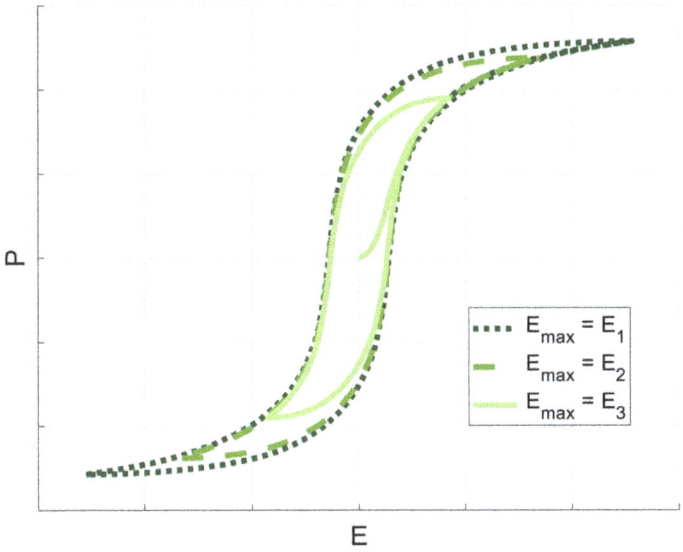

Figure 4.3. Simulated centred cycles (E_{max} is the maximum applied electric field).

In order to model the dielectric material behaviour precisely, it is necessary to accurately determine the distribution function from experimental data. There are mainly two options to determine it. The first assumes that the distribution function has a particular form (Lorentzian, Gaussian) and then determines the parameters of the chosen function to depict the average hysteretic behaviour. The second discretizes the distribution function in a finite set of values which are determined by suitable experimental data. In this chapter, the second option, which is expected to provide better accuracy, is chosen.

Two techniques have been tested for the acquisition of the discretized distribution function, namely, the centred cycle technique as described by Peng *et al* (2018) and by the Biorci's method (Biorci and Pescetti 1958). Both techniques provide relatively correct simulation results. A higher number of experimental data are required by the first technique but a correct behaviour can be reached with a lower size of the discretized distribution implementation and, this means a simpler and reduced memory management. Figure 4.4 gives an illustration of a Preisach's model distribution triangle obtained for a soft PZT ceramic.

For the extension to the dynamic behaviour, the Preisach's model is frequently used in its inverse configuration, i.e. working with P as the input to the model. To limit numerical issues in the model parameter, the Preisach's distribution used is still the direct $P(E)$'s distribution. For a given step time (t), $P(t)$ comes from the excitation waveform. If $\mathrm{d}P/\mathrm{d}t(t)$ is positive, an 'E high limit' (E_{high}) is determined on the Preisach's triangle. It corresponds to the E_i discretized step in the increasing direction which reflects a value of P just superior to $P(t)$. The 'E low limit' (E_{low}) is given by the E_{i-1} discretized step just beneath which is related to a P value just inferior to $P(t)$. Once the corresponding P_{low} and P_{high} limits linked to E_{low} and E_{high} limits, respectively, are calculated, it is assumed that P will vary linearly from its

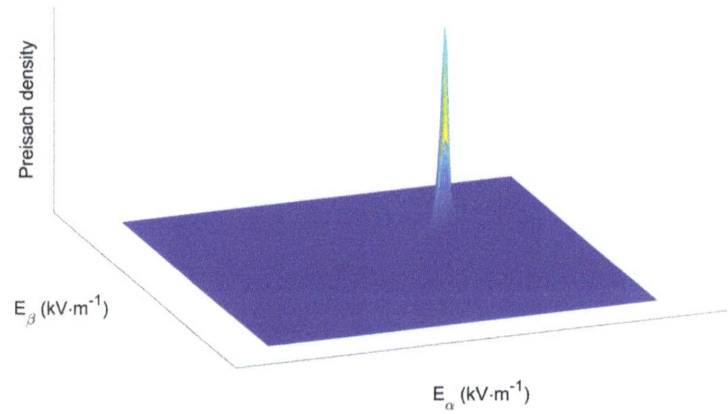

Figure 4.4. Preisach's model distribution triangle for a typical soft PZT ceramic.

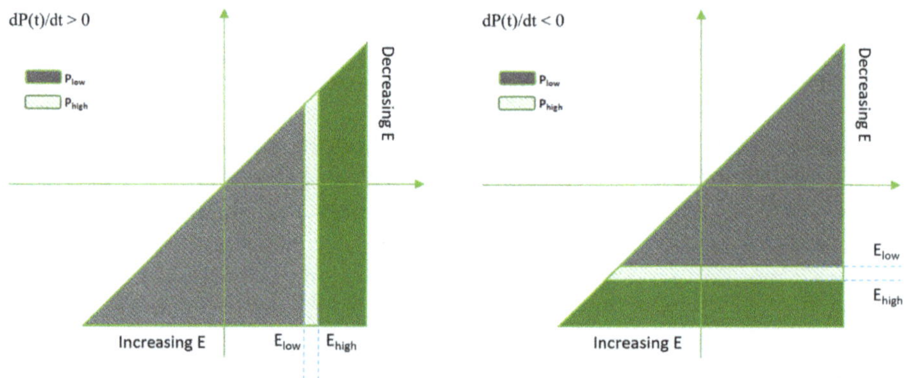

Figure 4.5. Illustration of the inverse Preisach's model: increasing (left side) and decreasing (right side) polarization (*P*).

inferior '*P* low limit' to its superior value '*P* high limit'. A precise *P*(*t*) is calculated thanks to a linear interpolation:

$$E(t) = E_{\text{low}} + (P(t) - P_{\text{low}}) \cdot \frac{E_{\text{high}} - E_{\text{low}}}{P_{\text{high}} - P_{\text{low}}}$$

If d*P*/d*t*(*t*) is negative, the same protocol, i.e. the determination of '*E* high and low limits', is used but in this scenario, *E* is read on the decreasing direction of the Preisach's triangle (figure 4.5).

The Jiles–Atherton model

Same as the Preisach's model, the Jiles–Atherton (J–A) model has been originally used to describe the hysteresis phenomenon in ferromagnetic materials (Jiles and Atherton 1984, 1986). The J–A model shows interesting properties, including:

- It requires fairly small memory storage, where the evolution history information is limited to a one-step time before the value of *P*;

- The P variations are completely described by a set of five parameters;
- The five parameters are related to microstructural properties, allowing physical interpretation of the J–A model;
- The model is easily reversible.

In the ferroelectric version of the J–A model, P is decomposed into its reversible (P_{rev}) and irreversible (P_{irr}) components (Leite *et al* 2003). These two contributions are linked through the following relation:

$$P_{rev} = \varsigma(P_{anh} - P_{irr})$$

where the anhysteretic polarization (P_{anh}) and the effective field ($E_e = E + \gamma \cdot P$) are linked via a sigmoid function, such as the Langevin function:

$$P_{anh} = P_{sat}\left(\coth\left(\frac{E_e}{a}\right) - \frac{a}{E_e}\right)$$

with

$$\frac{dP_{irr}}{dE_e} = \frac{P_{anh} - P_{irr}}{\kappa\delta}$$

P_{sat}, a, γ, ς and κ are the model parameters which can be obtained from measured hysteresis loops: P_{sat}-saturation polarization of the material, α-quantifies interdomain coupling in the dielectric material, a-quantifies domain wall density in the dielectric material, c-polarization reversibility, k-quantifies the average energy required to break the pinning site in the dielectric material. δ is a directional parameter assuming the value +1 if $dE(t)/dt > 0$ and −1 if $dE(t)/dt < 0$. Combining the equations above the main equation of the original J–A model can be written as:

$$\frac{dP}{dE} = \frac{(1-\varsigma)\frac{dP_{irr}}{dE_e} + \varsigma\frac{dP_{anh}}{dE_e}}{1 - \gamma(1-\varsigma)\frac{dP_{irr}}{dE_e} - \gamma\varsigma\frac{dP_{anh}}{dE_e}} \tag{4.1}$$

The differential equation (4.1) allows one to calculate P with respect to E variations. However, in most applications of this chapter P is known before E. Li *et al* (2001) described in detail an inverse model for ferroelectric materials where the model excitation is P. The physical principles of the model are similar and the main equation of this inverse model for ferroelectric materials can be:

$$\frac{dP}{dD} = \frac{(1-\varsigma)\frac{dP_{irr}}{dD_e} + \varsigma\frac{dP_{anh}}{dE_e}}{1 - \varepsilon_0(1-\gamma)(1-\varsigma)\frac{dP_{irr}}{dD_e} - \varsigma(1-\gamma)\frac{dP_{anh}}{dE_e}}$$

where

$$\frac{DP_{irr}}{dD_e} = \frac{P_{anh} - P_{irr}}{\varepsilon_0\kappa\delta}$$

D is the electrical polarization and $D_e = \varepsilon_0 \cdot E_e + P$ the effective electrical polarization (where ε_0 is the vacuum permittivity).

The dry-friction model

The dry-friction (DF) model assumes that only thermodynamic metastable states give rise to the anhysteretic curve $P = P_{anh}(E)$, and the presence of potential wells generate sequences of small discontinuous jumps similar to what is observed under mechanical dry-friction conditions. Under these conditions, the resultant polarization behaviour produces a well-defined hysteresis response as the number of discontinuous jumps approaches infinity.

The construction of the quasi-static DF model is based on a major $P(E)$ hysteresis cycle obtained by applying the following sequence of relationships between $E(t)$ and $P(t)$ as a function of time (t) for a single DF element.

Sequence equation (4.2):

$$\begin{gathered}\text{If } E(t) > f^{-1}(P(t - \mathrm{d}t)) + E_c \\ P(t) = f(E(t) - E_c) \\ \text{If } E(t) < f^{-1}(P(t - \mathrm{d}t)) - E_c \\ P(t) = f(E(t) + E_c) \\ \text{Else } P(t) = P(t - \mathrm{d}t)\end{gathered} \tag{4.2}$$

E_c is the coercive electrical field and $f()$ is a sigmoid function related to the anhysteretic behaviour, which is denoted as $P(E)$ and is defined as follows:

$$P_{anh} = \frac{2}{\pi} P_{sat} \tan^{-1}\left(\frac{E}{a}\right)$$

where P_{sat} and a are the anhysteretic function parameters. Sequence equation (4.2) above provides a good approximation for the main features of a major hysteresis cycle observed under steady-state conditions with symmetrical high-amplitude E excitation. However, this sequence is incapable of modelling transient phases, such as the first polarization curve and minor loops, although this restriction can be mitigated. Guyomar *et al* (2007, 2009) and Ducharne *et al* (2013) addressed this issue by replacing the single DF element with a system composed of q elements having different coercive electrical field values and different weights characterizing their individual impacts on the overall hysteresis behaviour. The distribution of the coercivity values of the elements was considered homogeneous with a constant ΔE step, and their corresponding weights were defined according to the following spectrum function:

$$P = \sum_{i=1}^{q} \mathrm{Spectrum}(E_{c_i}) P_i$$

Good results can be obtained by defining the spectrum function according to the following Gaussian function:

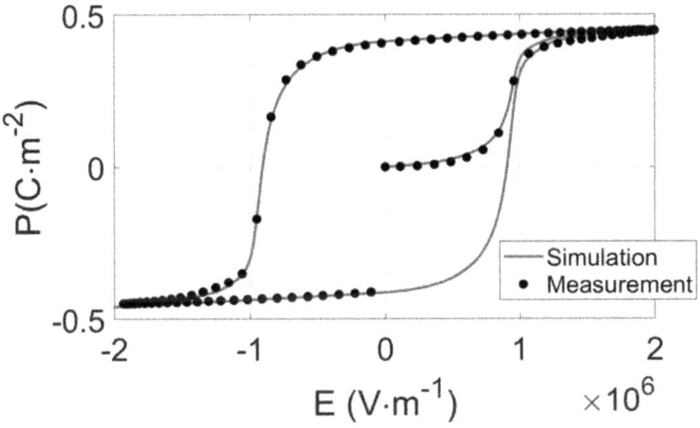

Figure 4.6. Comparison of simulation and measurement for the major hysteresis cycle with the DF model for a typical soft PZT ceramic.

$$\text{Spectrum}(E_{c_i}) = \frac{e^{-\frac{(E_{c_i}-\varsigma)^2}{2\sigma^2}}}{\sigma\sqrt{2\pi}}$$

where ς is the Gaussian function average and σ is its standard deviation. The product $\varsigma \cdot \Delta E$ is equal to the coercivity of the simulated hysteresis cycle $P(E)$. Similarly, σ is defined according to the ratio of the differential permittivities of the simulated anhysteretic curve $P(E)$ and at the coercivities of the simulated $P(E)$ curve. The use of a Gaussian function is particularly beneficial because it involves only two parameters ς and σ. These parameters can be set easily based on experimental data. However, a previous study adopted a numerical solution for defining the spectrum function by deconvolving the first polarization curve by the anhysteretic curve (Guyomar *et al* 2007). While this solution produced very accurate results, the methodology requires the anhysteretic curve which is difficult to obtain experimentally, and the complexity of calculating the spectrum curve increases exponentially with an increasing number of DF elements because each DF element is characterized by its own parameter. In contrast to it, the use of a Gaussian function with only two parameters means that the number of DF elements in the DF model is not constrained. Finally, the distribution of the spectrum function must be normalized such that:

$$\sum_{i=1}^{q} \text{Spectrum}(E_{c_i}) = 1$$

Figure 4.6 illustrates the DF model accuracy by comparing experimental and simulated first polarization curves for the major hysteresis loop of a typical PZT.

4.3.2 Dynamic contribution: extension of the quasi-static contribution to the dynamic behaviour

Under weak frequency conditions, the scalar quasi-static lump hysteresis models described above provide accurate results for the evolution of P–E loops. However, as soon as the quasi-static external conditions become invalid, significant errors occur. Improvements can be made by adding a simple dynamic contribution to the lump model, a damping term proportional to the time derivative of P. This term is homogeneous to an equivalent excitation field but for this simple approach, the correct simulation results are limited to a narrow frequency bandwidth. It seems that a simple viscous loss term $\mathrm{d}P/\mathrm{d}t$ leads to an overestimation in the high frequency part of the hysteresis loop area versus the frequency curve. Therefore, another correction of the lump model must be introduced to obtain correct simulation results over a wide frequency bandwidth. A mathematical operator dealing with the low frequency and the high frequency components in a different way than a straight time derivative is required. Such operators can be found in the framework of fractional calculus. They are the so-called non-integer derivatives or fractional derivatives.

Fractional time derivative $\mathrm{d}^nP/\mathrm{d}t^n$, where n is a real positive number, can be added to the lump model as referred to by Grünwald–Letnikov or Riemman–Liouville definitions (Grünwald 1867, Riemann 1892, Ortigueira 2011, Liouville 1832). Both are cases of a general fractional order operator. The former represents the nth order derivative, while the latter represents the mth fold integral (for negative, $m = -n$). In this sense, the class of functions described by the Riemman–Liouville definition is broader (the function must be integrable) than the one defined by Grunwald and Letnikov. However, for a function from the Grunwald–Letnikov class, both definitions are equivalent. In this chapter, the Riemman–Liouville form is used for $n \in [0,1]$ and is expressed by equation (4.3).

$$\frac{D^n f(t)}{\mathrm{d}t^n} = D_t^n f(t) = \frac{1}{\Gamma(1-n)}\frac{\mathrm{d}}{\mathrm{d}t}\int_{-\infty}^{t}(t-\tau)^{-n} f(t)\mathrm{d}\tau \tag{4.3}$$

Γ is the Euler gamma function and τ is a dummy variable. According to equation (4.3), the fractional derivative of an arbitrary integrable function $f(t)$ can also be considered as the convolution of a $f(t)$ function and $t^n/\Gamma(1-n)$. In this case, n is the order of the fractional derivation. The additional time derivative present in the formula coincides with the occurrence of positive argument of the gamma function, $\Gamma()$, leading its convergence to a finite value. It is obvious in equation (4.3) that the fractional derivative includes memory of the previous states. From a spectral point of view, an interesting property of the fractional calculus is that in the frequency domain, the frequency spectrum $f(\omega)$ of $f(t)$ is multiplied by $(j\omega)^n$ instead of $j\omega$ for classic first-order derivation (Guyomar *et al* 2008, 2010a, Ortigueira 2011). As a consequence, a new degree of freedom (n value) is available and by adjusting it, one can satisfy the experimental balance requirements, i.e. evolution of the hysteresis area versus the frequency curve.

The fractional derivative is introduced into the lumped quasi-static hysteresis model through a dynamic contribution (figure 4.7). The term $\rho\cdot\mathrm{d}P/\mathrm{d}t$ is replaced by $\rho\cdot\mathrm{d}^nP/\mathrm{d}t^n$. This contribution is then added to the quasi-static contribution as shown in equation (4.4) (where η is a constant).

$$\eta\cdot\frac{\mathrm{d}^nP(t)}{\mathrm{d}t^n} = E_{\mathrm{dyn}}(t) - f_{\mathrm{static}}^{-1}(P(t)) \tag{4.4}$$

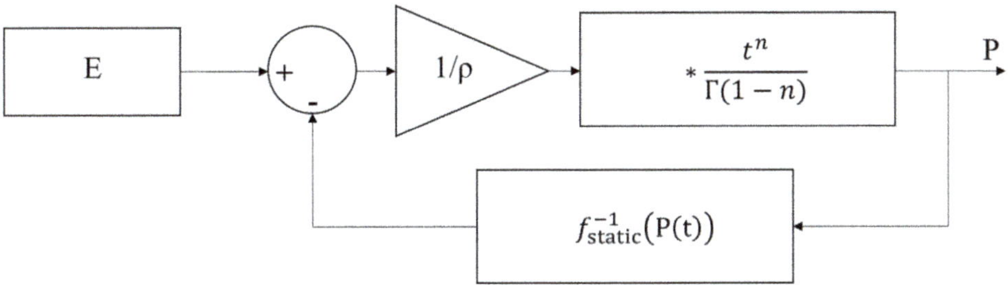

Figure 4.7. Illustration of the ferroelectric hysteresis model including the dynamical contributions.

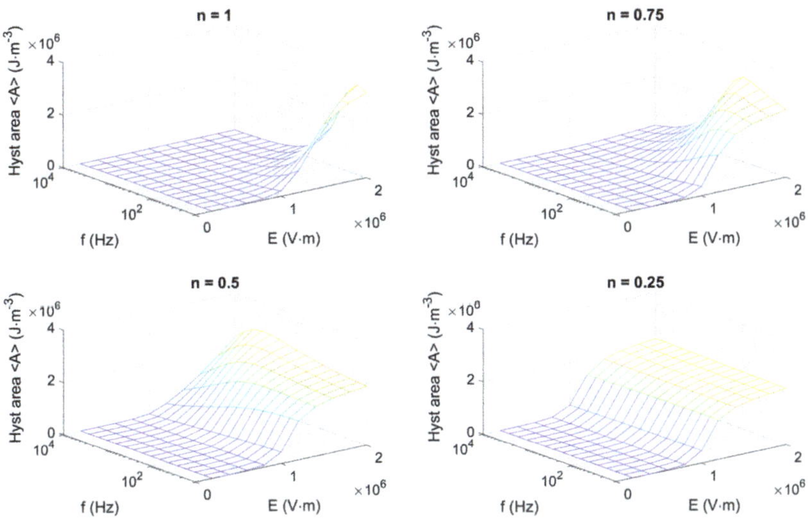

Figure 4.8. Bloc-scheme illustration for the ferroelectric hysteresis model, including the dynamical contribution.

Illustration of the frequency effect and simulation method efficiency can be found in figure 4.8 and elsewhere (Guyomar *et al* 2007, 2009, Ducharne *et al* 2013).

4.3.3 Temperature effect

The effect of temperature on the dielectric properties of classic ferroelectric materials is complex, especially when phase changes occur. The phenomenological simulation methods described above cannot deal with these changes. This limits the validity of the simulation methods to the phase conservation condition. The dielectric properties remain largely unchanged for slight variations from room temperature. For larger temperature variations in the upward direction, a continuous decrease of both the saturation polarization and the coercive field is observed in many ferroelectric materials. Some ferroelectric materials exhibit an increased hysteresis towards a maximum level, followed by a similar decrease towards a fully vanished loop at the Curie temperature.

Ferroelectric to non-ferroelectric phase transition occurs at the Curie temperature, which thoroughly modifies the properties. The variations can be observed on the hysteresis cycle area $<A>$ versus the temperature curve $<A>(\theta)$.

Raghunathan *et al* (2009) proposed to model the temperature dependence of ferromagnetic hysteresis based on the J–A theory. Recently, Bai *et al* (2018) and Ducharne *et al* (2020) described a similar approach for the temperature dependence of ferroelectric materials. This approach aims to assign a temperature dependence to the five J–A model parameters. Based on Raghunathan *et al*'s work, the five temperature-dependent parameters become:

$$P_{\text{sat}}(T) = P_{\text{sat}}(0) \cdot \left(1 - \frac{T}{T_{\text{c}}}\right)^{\chi}$$

$$\kappa(T) = \kappa(0) \cdot \mathrm{e}^{\frac{-2T}{\chi \cdot T_{\text{c}}}}$$

$$a(T) = a(0) \cdot \left(1 - \frac{T}{T_{\text{c}}}\right)^{\chi}$$

$$\gamma(T) = \gamma(0) \cdot \left(1 - \frac{T}{T_{\text{c}}}\right)^{\chi}$$

$$\varsigma(T) = \varsigma(0) \cdot \left(1 - \frac{T}{T_{\text{c}}}\right)^{\chi}$$

Here 0 stands for 0 K, for instance, $P_{\text{sat}}(0)$ is the saturation polarization at 0 K. T is temperature and T_{C} is Curie temperature. As it is complicated to obtain experimental data at 0 K, one solution is to consider $P_{\text{sat}}(0)$, $k(0)$, $a(0)$, $\alpha(0)$ and $c(0)$ to be equal to their room temperature values, assuming their variations in the decreasing direction are null. χ is a material-dependent critical exponent (Raghunathan *et al* 2009).

4.3.4 Light effect

When subject to photon energy higher than the band gap, the properties of ferroelectric materials may also be influenced by incident light (Paillard *et al* 2016). Although it is well known that ferroelectrics are sensitive to thermal, electric and mechanical stimuli, the light dependence has been much less studied in simulations.

For some families of ferroelectrics, the experimental variations of the hysteresis loop shape versus different lighting conditions show strong correlations. All the hysteresis indicators, including absolute values of coercive field, saturation polarization and remanent polarization, may show increasing values with an increase of the light power. A physical interpretation of these changes can be the simultaneous creation of free charges and surface dielectric dipoles at a microscopic scale. Light will affect the interfacial polarization and consequently the related domain wall motion, domain displacements and their configuration. The variations of the hysteresis loop shape as a function of light are close to those of temperature. One possible simulation approach relies on the same mechanism. A scaling relation can even be proposed between light and temperature. Similar to the situation of temperature dependence, a modified light-dependent J–A model adds lighting factors to its five parameters:

$$P_{\mathrm{sat}}(\mathrm{LPD}) = P_{\mathrm{sat_d}}(0) \cdot \left(1 - \frac{\mathrm{LPD}}{\mathrm{LPD_{max}}}\right)^{\iota}$$

$$\kappa(\mathrm{LPD}) = \kappa_{\mathrm{d}} \cdot \mathrm{e}^{\frac{-2\mathrm{LPD}}{\iota \cdot \mathrm{LPD_{max}}}}$$

$$a(\mathrm{LPD}) = a_{\mathrm{d}} \cdot \left(1 - \frac{\mathrm{LPD}}{\mathrm{LPD_{max}}}\right)^{\iota}$$

$$\gamma(\mathrm{LPD}) = \gamma_{\mathrm{d}} \cdot \left(1 - \frac{\mathrm{LPD}}{\mathrm{LPD_{max}}}\right)^{\iota}$$

$$\varsigma(\mathrm{LPD}) = \varsigma_{\mathrm{d}} \cdot \left(1 - \frac{\mathrm{LPD}}{\mathrm{LPD_{max}}}\right)^{\iota}$$

where LPD stands for light power density and $\mathrm{LPD_{max}}$ corresponds to a maximum of this power density. In the resolution, $\mathrm{LPD_{max}}$ takes the form of a notional value obtained from comparisons between simulations and measurements and is set through an optimization method. $P_{\mathrm{sat_d}}$, k_{d}, a_{d}, α_{d} and c_{d} are the simulation parameters set under dark conditions. ι is a material critical exponent.

Improved simulation results can be obtained by considering a leakage current, especially under high values of LPD. One can take into account this leakage current by adding an electric field-dependent contribution:

$$P(t) = \int \frac{\mathrm{d}P}{\mathrm{d}E}(t) \cdot \mathrm{d}E + \frac{1}{\rho_0} \int E(t) \cdot \mathrm{d}t$$

$$\rho_0 = \frac{\varepsilon}{\vartheta}$$

where ϑ is the electric conductivity and ε is a constant depending on the sample geometry. Additional information and data such as comparisons between the simulations and measurements are given by Ducharne *et al* (2020).

4.3.5 Stress effect

It is known that P decreases under external compressive stress. This reduction degrades the piezoelectric properties due to the ferroelectric–ferroelastic field readjustments. On one hand, for symmetry of the ferroelectric behaviour, an electric field generates a ΔP variation independent of the sign of dP/dt. On the other hand, an increase of stress leads to a ΔP variation opposite to the sign of P. Consequently, the symmetry of P/$-P$ breaks down for an electric field stimulus (vector quantity) but is retained for stress (tensor quantity). Thus, stress can be considered in the model as an equivalent electric field: F (applied force) = E or $h(P)\cdot T$, where $h(P)$ induces the symmetry described above. $h(P)$ can take the form of any odd function, with the simplest being $h(P) = \alpha_0\cdot P$. The association between mechanical and electrical excitations is confined in $E_{equ} = \alpha_0\cdot T\cdot P$ where α_0 is a material characteristic. This relation is independent of the sample geometry and is defined through an optimization process. This ascending association amid electric and mechanical parameters enables the simulation of the ceramic component regardless of the nature of the excitation stress.

Such a straightforward approach gives satisfactory simulation results, as illustrated by Guyomar *et al* (2010a) and Zhang *et al* (2013, 2018). It should be noted that multi-axiality of the stress is not considered here.

4.4 Summary

In this chapter, three quasi-static ferroelectric hysteresis simulation methods have been introduced. The quasi-static contribution is the key element of a phenomenological hysteresis model. It depends on the material's history and it deserves a precise and sophisticated description.

A dynamic extension based on fractional derivative operators has been discussed. Classic first-order derivative contributions are limited. Switching to fractional derivatives provides flexibility in the simulation methods and offers accurate results in broad frequency windows from very low-frequency ranges (Ducharne *et al* 2019) to very large ranges (Guyomar *et al* 2010b).

The dependence of ferroelectric behaviour on temperature and light is similar with the same mechanism used in their simulations. The mechanism is based on a modified J–A model having its five parameters dependent on temperature and light.

The influence of stress can be reflected by a simple method which relies on experimental observations and can easily be implemented for all the quasi-static contributions discussed in this chapter.

The phenomenological aspect of the simulation methods described in this chapter limits the predictivity to some well-defined experimental situations with phase transitions being absent. If such conditions are respected, the methods can be applied successfully in predicting the functionality of ferroelectrics working under different stimuli.

References

Arlt G 1996 A physical model for hysteresis curves of ferroelectric ceramics *Ferroelectrics* **189** 103–19

Bai Y, Ducharne B, Jantunen H and Juuti J 2018 Simulation and validation of temperature-dependent ferroelectric properties of multifunctional BCZT and KNBNNO ceramics *Mater. Res. Express* **5** 116305

Bernard Y, Maalej A H, Lebrun L and Ducharne B 2007 Preisach modelling of ferroelectric behaviour *Int. J. Appl. Electrom* **25** 729–33

Biorci G and Pescetti D 1958 Analytical theory of the behaviour of ferromagnetic materials *Il Nuovo Cimento (1955–1965)* **7** 829–42

Bowen C R, Kim H A, Weaver P M and Dunn S 2014 Piezoelectric and ferroelectric materials and structures for energy harvesting applications *Energ. Environ. Sci.* **7** 25–44

Damjanovic D 2006 *Hysteresis in Piezoelectric and Ferroelectric Materials* (Lausanne: Academic) pp 337–465

Ducharne B, Guyomar D and Sebald G 2007 Low frequency modelling of hysteresis behaviour and dielectric permittivity in ferroelectric ceramics under electric field *J. Phys. D: Appl. Phys.* **40** 551

Ducharne B, Juuti J and Bai Y 2020 A simulation model for narrow band gap ferroelectric materials *Adv. Theor. Sim.* **3** 2000052

Ducharne B, Newell B and Sebald G 2019 A unique fractional derivative operator to simulate all dynamic piezoceramic dielectric manifestations: from aging to frequency-dependent hysteresis *IEEE T. Ultrason. Ferr.* **67** 197–206

Ducharne B, Zhang B, Zhang B and Sebald G 2013 Fractional derivative operators for modeling piezoceramic polarization behaviors under dynamic mechanical stress excitation *Sensors Actuators* A **189** 74–9

Grünwald A K 1867 Über 'begrenzte' Derivation und deren Anwendung *Z. Angew. Math. Phys* **12** 441–80

Guyomar D, Ducharne B and Sebald G 2008 Time fractional derivatives for voltage creep in ferroelectric materials: theory and experiment *J. Phys. D: Appl. Phys.* **41** 125410

Guyomar D 2010a High frequency bandwidth polarization and strain control using a fractional derivative inverse model *Smart Mater. Struct.* **19** 045010

Guyomar D 2010b The use of fractional derivation in modeling ferroelectric dynamic hysteresis behavior over large frequency bandwidth *J. Appl. Phys.* **107** 114108

Guyomar D, Ducharne B and Sebald G 2007 Dynamical hysteresis model of ferroelectric ceramics under electric field using fractional derivatives *J. Phys. D: Appl. Phys.* **40** 6048

Guyomar D, Ducharne B, Sebald G and Audiger D 2009 Fractional derivative operators for modeling the dynamic polarization behavior as a function of frequency and electric field amplitude *IEEE T. Ultrason. Ferr.* **56** 437–43

Hegewald T, Kaltenbacher B, Kaltenbacher M and Lerch R 2008 Efficient modeling of ferroelectric behavior for the analysis of piezo-ceramic actuators *J. Intel. Mat. Syst. Str* **19** 1117–29

Hu H and Mrad R B 2003 On the classical Preisach model for hysteresis in piezo-ceramic actuators *Mechatronics* **13** 85–94

Hughes D and Wen J T 1997 Preisach modeling of piezoceramic and shape memory alloy hysteresis *Smart Mater. Struct.* **6** 287

Jiles D C and Atherton D L 1984 Theory of ferromagnetic hysteresis *J. Appl. Phys.* **55** 2115–20
Jiles D C and Atherton D L 1986 Theory of ferromagnetic hysteresis *J. Magn. Magn. Mater.* **61** 48–60

Kamlah M J C M 2001 Ferroelectric and ferroelastic piezo-ceramics modeling of electromechanical hysteresis phenomena *Continuum Mech. Therm.* **4** 219–68

Klinkel S 2006 A phenomenological constitutive model for ferroelastic and ferroelectric hysteresis effects in ferroelectric ceramics *Int. J. Solids Struct.* **43** 7197–222

Krueger H H 1968 Stress sensitivity of piezoelectric ceramics: part 3. Sensitivity to compressive stress perpendicular to the polar axis *J. Acoust. Soc. Am.* **43** 583–91

Leite J V, Sadowski N, Kuo-Peng P, Batistela N J and Bastos J P A 2003 The inverse Jiles–Atherton model parameters identification *IEEE T. Magn.* **39** 1397–400

Li Y L, Hu S Y, Liu Z K and Chen L Q 2001 Phase-field model of domain structures in ferroelectric thin films *Appl. Phys. Lett.* **78** 3878–80
Liouville J 1832 Mémoire sur quelques questions de géométrie et de mécanique et sur un nouveau genre de calcul pour résoudre ces équations *Ecole Polytechnique* **13** 71–162
Lynch C S 1996 The effect of uniaxial stress on the electro-mechanical response of 8/65/35 PLZT *Acta Mater.* **44** 4137–48
Martin R M 1972 Piezoelectricity *Phys. Rev.* B **5** 1607
Masui S, Yokozeki W, Oura M, Ninomiya T, Mukaida K, Takayama Y and Teramoto T 2003 Design and applications of ferroelectric nonvolatile SRAM and flip-flop with unlimited read/program cycles and stable recall *IEEE Proc. 2003 Custom Integrated Circuits Conf. (San Jose, CA, September 2003)* (Piscataway, NJ: IEEE) pp 403–6
Mayergoyz I 1986 Mathematical models of hysteresis *IEEE T. Magn.* **22** 603–8
Ortigueira M D 2011 *Fractional Calculus for Scientists and Engineers* vol 84 (Berlin: Springer)
Paillard C, Bai X, Infante I C, Guennou M, Geneste G, Alexe M, Kreisel J and Dkhil B 2016 Photovoltaics with ferroelectrics: current status and beyond *Adv. Mater.* **28** 5153–68
Peng D, Sima W, Yang M, Zou M, Zhou Y and Liu Y 2018 An improved centered cycle method for identifying the Preisach distribution function *IEEE T. Magn.* **54** 1–5
Picinin A, Lente M H, Eiras J A and Rino J P 2004 Theoretical and experimental investigations of polarization switching in ferroelectric materials *Phys. Rev.* B **69** 064117
Preisach F 1935 Über die magnetische Nachwirkung *Z. Phy* **94** 277
Raghunathan A, Melikhov Y, Snyder J E and Jiles D C 2009 Modeling the temperature dependence of hysteresis based on Jiles–Atherton theory *IEEE T. Magn.* **45** 3954–7
Riemann B 1892 *Gesammelte Mathematische Werke und Wissenschaftlicher Nachlass* 2nd edn (Leipzig: Teubner) eds H Weber and R Dedekind
Smith R C, Seelecke S, Ounaies Z and Smith J 2003 A free energy model for hysteresis in ferroelectric materials *J. Intel. Mat. Syst. Str* **14** 719–39
Sutor A, Rupitsch S J and Lerch R 2010 A Preisach-based hysteresis model for magnetic and ferroelectric hysteresis *Appl. Phys.* A **100** 425–30
Wang D, Fotinich Y and Carman G P 1998 Influence of temperature on the electromechanical and fatigue behavior of piezoelectric ceramics *J. Appl. Phys.* **83** 5342–50
Wang L, Wang X and Shi J 2010 Measurement and estimation of ferroelectric hysteresis loops *Ferroelectrics* **411** 86–92
Wang T H, Tu C S, Ding Y, Lin T C, Ku C S, Yang W C, Yu H H, Wu K T, Yao Y D and Lee H Y 2011 Phase transition and ferroelectric properties of xBiFeO$_3$–(1−x)BaTiO$_3$ ceramics *Curr. Appl Phys.* **11** S240–3
Weaver P M, Cain M G, Correia T M and Stewart M 2011 Electromechanical coupling and temperature-dependent polarization reversal in piezoelectric ceramics *IEEE Trans. Ultrason. Ferr.* **58** 1730–6
Xu Y 2013 *Ferroelectric Materials and Their Applications* (Los Angeles, CA: Elsevier)
Zhang B, Ducharne B, Gupta B, Sebald G, Guyomar D and Gao J 2018 Experimental sea wave energy extractor based on piezoelectric Ericsson cycles *J. Intel. Mat. Syst. Str* **29** 1102–12
Zhang B, Ducharne B, Guyomar D and Sebald G 2013 Energy harvesting based on piezoelectric Ericsson cycles in a piezoceramic material *Eur. Phys. J-Spec. Top.* **222** 1733–43

Part II

Emerging ferroelectrics and their novel applications

Chapter 5

Photoferroelectrics and energy harvesting

Yang Bai

This chapter extends the theoretical discussions of part I to experimental studies and applications. The interplay between ferroelectric properties and stimuli of stress, temperature and electric field is relatively well-known from the experimental aspect and has seen continuous development of appropriate theoretical methods for its understanding (as discussed in chapters 2–4). The influence of another found stimulus, light, still requires more experimental work to support the theory development. This chapter discusses milestone experimental studies on light–ferroelectric interaction and subsequently defines the modern photoferroelectrics, eventually proceeding towards application-oriented topics, such as band gap engineering for opto-ferroelectric and ferroelectric photovoltaic devices, and multi-energy harvesters/sensors based on photoferroelectrics. The chapter also includes preliminary discussions on domain walls and halide perovskites that will be discussed in detail in chapters 6 and 7, respectively.

5.1 History and rationale of light–ferroelectric interaction

As has been introduced in chapter 1, conventional research on ferroelectrics does not necessarily involve the interaction of light and domain wall motion. Light as an external stimulus to trigger a response by a ferroelectric was not studied until the 1970s (Glass *et al* 1974, Glass *et al* 1975, Chanussot and Glass 1976, Fridkin 1979) when the phenomenon of the photovoltaic effect observed for classic ferroelectrics exposed to incident light started to be discussed. This phenomenon is nowadays known as the bulk photovoltaic effect (BPVE) (Lopez-Varo *et al* 2016).

Subsequent research has paid particular attention to the BPVE and consequently a more comprehensive understanding has been established for both the BPVE and a more general physical/material behaviour which is the ferroelectric or polarization-assisted photovoltaic effect (FePV). For instance, the signature of the BPVE is its response and sensitivity to polarized light. As shown in figure 5.1 (Bhatnagar *et al* 2013), the angle between the vector of the polarized light and the direction of the

doi:10.1088/978-0-7503-3975-9ch5

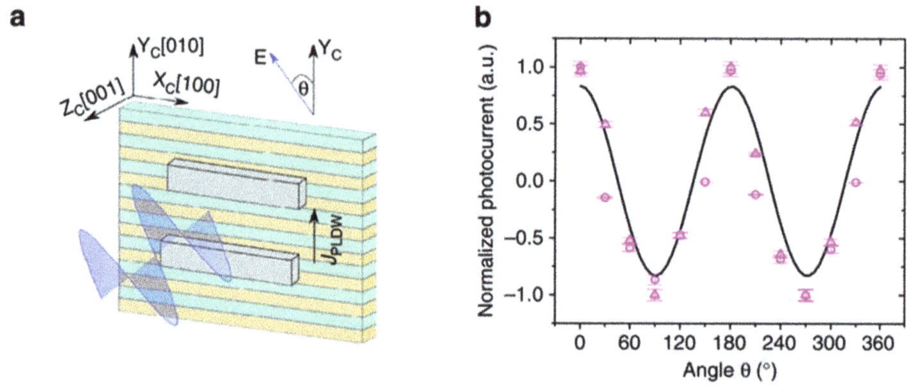

Figure 5.1. (a) Schematic of a $BiFeO_3$ thin film sample where the strips represent the 109° or 71° domain system. On top of the film is a pair of in-plane electrodes from which the photocurrent was extracted. The angle between the vector E of the linear polarized light and the direction of the current flow J_{PLDW} is defined as θ. (b) Dependence of normalized photocurrent on θ with the triangle data points representing the 109° domain system, the circle data points representing the 71° domain system, and the solid line representing the theoretical fitting. Reproduced under the Creative Commons CC-BY-NC-ND license from (Bhatnagar *et al* 2013).

photocurrent has a straightforward effect on the normalized photocurrent. The maxima are reached whenever the light polarization direction is in parallel with or orthogonal to the photocurrent direction.

The sensitivity to light polarization can be well explained by equation (5.1) (Fridkin 1979),

$$I_i = \Lambda \beta_{ijk} e_j e_k \tag{5.1}$$

where I_i is the photocurrent, Λ is the intensity of light, and β_{ijk} is a third rank BPVE tensor. In theory, the photocurrent in terms of its sign and magnitude depends on the domain (or crystal) orientation with respect to the projections of the electric field of linearly polarized light onto the plane of the sample along two orthogonal directions, e_j and e_k, respectively.

Meanwhile, under unpolarized light, the effectiveness of photocurrent extraction also depends on the angle between the electrode pair and the domain or crystal orientation. Figure 5.2 (Yang *et al* 2010) explains such a phenomenon that when the photocurrent direction is in parallel with the net ferroelectric polarization direction, a short-circuit photocurrent as well as an open-circuit photovoltage can be obtained (figures 5.2(a) and (c)). However, when the photocurrent direction is perpendicular to the ferroelectric polarization direction, the current–voltage curve goes through the zero point and thus no photovoltaic effect can be observed (figures 5.2(b) and (d)).

Despite being well-established experimentally, BPVE has not yet been used in practice in contrast to the semiconductor-based, interface-induced (junction) photovoltaics used in today's rapidly growing solar cell market. This is because classic ferroelectric compositions such as $BiFeO_3$, PZT ($Pb(Zr,Ti)O_3$) and $BaTiO_3$ based compounds have wide band gaps of >2.7 eV compared to 1.8–3.1 eV which is the

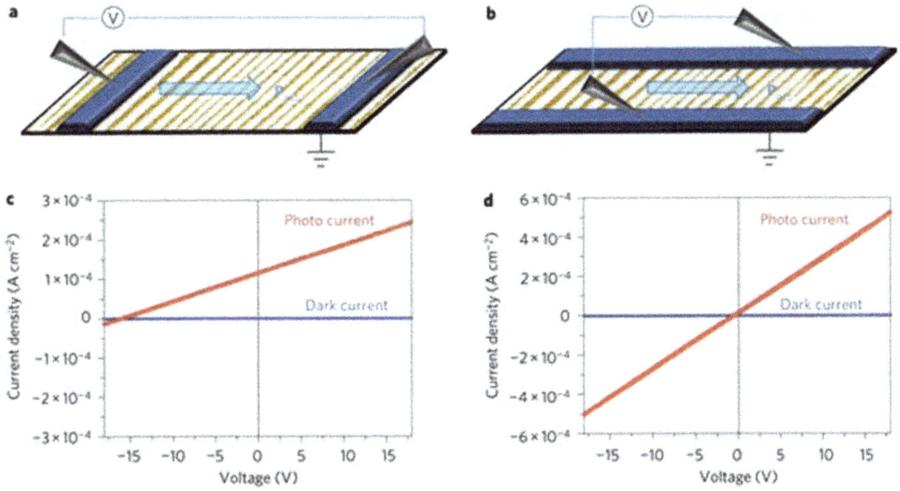

Figure 5.2. Schematics of $BiFeO_3$ thin film samples: (a) perpendicular and (b) parallel device geometries. (c)(d) Dependence of current density on bias voltage for the perpendicular and parallel devices, respectively. Reproduced from Yang *et al* (2010). Copyright © 2010 with permission of Springer Nature.

range of visible light in the solar radiation spectrum. The wide band gaps lead to the fact that over 90% of the solar energy cannot be absorbed and thus effectively converted to electricity (Bai *et al* 2019a). From the point of view of semiconductor physics, good ferroelectrics are typically strong insulators exhibiting extremely low conductivity (~10^{-9} S m^{-1} or even smaller). Their insulating nature helps them to sustain high electric fields and hence to induce a strong ferroelectricity. Nevertheless, it leads to a high effective mass of the charge carriers; additionally, the recombination rates of photo-excited charge carriers are high (Würfel and Würfel 2016). The poor light absorption and unfavourable recombination have resulted in negligible power conversion efficiencies (PCEs) that are on the order of 10^{-4} compared to on the order of 10^{-1} for conventional semiconductors.

Researchers have not been discouraged by the negligible PCEs of BPVE. Although its era of being used for standalone solar cells is yet to come, BPVE may indeed become an efficient addition or accessory to the configurations of existing solar cells due to its potential role of stimulating the separation of photo-generated charge carriers. Some works have pioneered the insertion of thin ferroelectric layers into the junction-based semiconductor solar cell structures (Pérez-Tomas *et al* 2019, Yang *et al* 2019). The polarization in the ferroelectric layers, as long as it points towards the direction of the electron flow, is likely to provide an extra potential by bending the bands at the ferroelectric–semiconductor interface. Consequently, more charge carriers are protected from being recombined and higher PCEs have been obtained. When used as an additional layer, it is crucial that the ferroelectric layer barely absorbs light (wide band gap) and is thin enough (maximum a few nm) to avoid inducing increased recombination.

BPVE and FePV have also intrigued researchers to link the interesting properties, such as long charge carrier lifetimes and the resultant high PCEs discovered in some halide perovskite compounds, to the possible presence of spontaneous polarization (Shahrokhi *et al* 2020), as will be discussed in chapter 7. The BPVE in oxide perovskites will be discussed in detail in section 5.2.

More recently, light has been found to be not only able to induce BPVE but also to result in domain wall movement (Rubio-Marcos *et al* 2018, Vats *et al* 2019). Figure 5.3 (Rubio-Marcos *et al* 2018) illustrates that when an incident light with wavelength of 532 nm is shone on a $BaTiO_3$ single crystal sample, the population of the *c* domain (polarization along the pre-defined *Z*-axis) increases. Observed directly under XRD (x-ray diffraction), this effect provides unambiguous evidence of light-induced domain wall motion. Indirect evidence, e.g. a change of permittivity when subject to incident light, has also been obtained in the same work.

The light-induced domain wall motion is a striking discovery. It means that light, or more generally radiation waves, will become another external stimulus in addition to electric field, stress and temperature that can manipulate domain structures. The light-induced domain wall motion will also be discussed in detail in section 5.2. This phenomenon opens opportunities for many fascinating applications, for example, using light as a virtual gate in ferroelectric field-effect transistors, non-volatile memories, artificial synapses for neuromorphic computing with

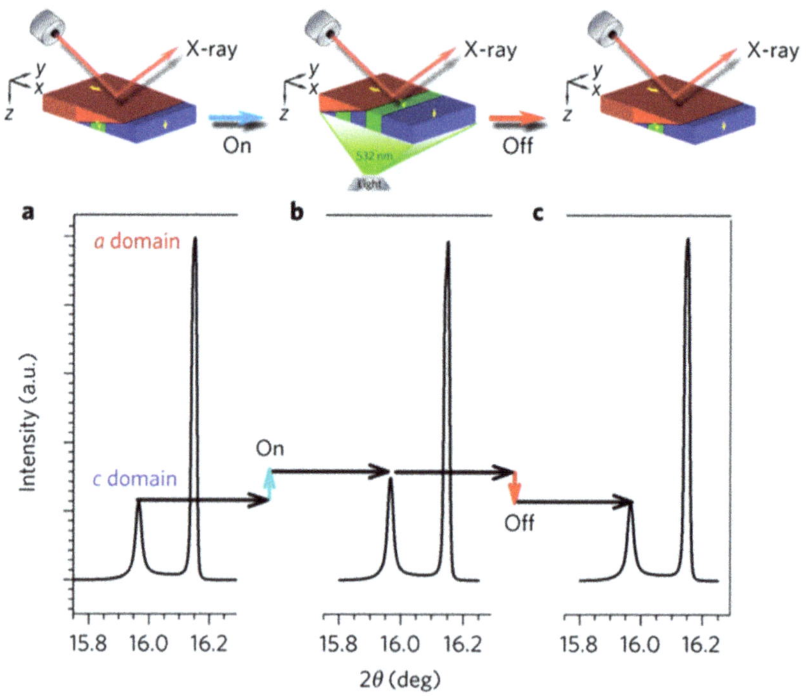

Figure 5.3. The schematics on the top show the setup of the XRD (incident x-ray on the top surface of the sample), incident light (on the bottom surface) and the sample configuration with different colours representing different domain orientations. (a)–(c) Evolution of the XRD patterns of *a* and *c* domains when the incident light is turned on and off. Reproduced from Rubio-Marcos *et al* (2018). Copyright © 2017 with permission from Springer Nature.

writing/reading functions, optoelectrical dual-source actuators for remote and precise control of micromachines and medical operations, etc (Cheng *et al* 2020, Fu *et al* 2020, Vats *et al* 2020, Bai 2021b, Rubio-Marcos *et al* 2021).

Beyond the pristine wide band gaps of standard ferroelectrics, a potentially tunable and reduced band gap, especially when extending deep into the visible-light range, will broaden the applications of the light–ferroelectric interaction. Band gap engineering of ferroelectrics will be discussed in section 5.3. The resultant visible-light photovoltaic effect arising owing to the reduced band gaps, and co-existing with the piezoelectric and pyroelectric effects, will enable single materials to become truly integrated, multi-functional sensors and energy harvesters responsive to multiple stimuli or energy sources (i.e. converting optical, mechanical and thermal energies simultaneously into electricity) (Bai *et al* 2017b, 2018, 2020). This will help to promote miniaturization of hybrid and multi-functional systems and will be discussed in section 5.4.

5.2 Solid-state physics from the ferroelectric aspect—what is photoferroelectrics?

With the rapid development of the research topic of light–ferroelectric interaction, comprehensively defining the associated term, photoferroelectrics, can be complex. In general, photoferroelectrics may exhibit electrical properties of both classic ferroelectrics and semiconductors. More importantly, as has been discussed in section 5.1, the incident light should be able to trigger domain wall motion in ferroelectrics, namely, 'photo'-'ferroelectric' effect. Vice versa, the presence of ferroelectric polarization should also induce a photovoltaic effect which is in other words a 'ferroelectric'-'photo'-related effect. Therefore, this section will discuss the definition and identification of modern photoferroelectrics from these three aspects.

5.2.1 Conductivity and band gap

The conductivity (σ) of a classic semiconductor can be defined by equation (5.2),

$$\sigma = 10^{25}e(\mu_e + \mu_h)\exp\left(-\frac{E_g}{2kT}\right) \tag{5.2}$$

where e is the electronic charge, μ_e and μ_h are electron and hole mobilities, respectively, E_g is the band gap of the material, k is the Boltzmann constant, and T is temperature (Moulson and Herbert 2003). Figure 5.4 compares the conductivities of a classic semiconductor, a classic insulating ferroelectric, and a band gap engineered (with reduced band gap) ferroelectric material.

Most semiconductors have relatively high $\mu_e + \mu_h$ and for those with low E_g such as Si, the conductivity usually reaches the level of $\geqslant 10^{-3}$ S m^{-1}. Such a large conductivity, even if spontaneous polarization exists in the material, is unlikely to give measurable ferroelectricity as any applied electric field will generate a significant current flowing through the material rather than being accumulated at the surfaces to trigger polarization switching. It will also be difficult to observe a drastic change

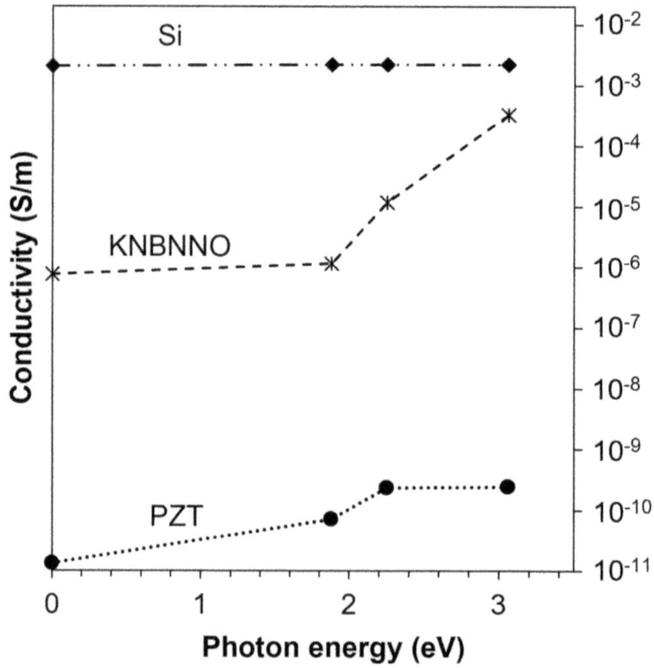

Figure 5.4. Dependence of conductivity on the photon energy of incident light for a p-type monocrystalline Si, a commercial PZT-5A ceramic, and a band gap engineered ferroelectric ceramic, Ba A-site doped Ni B-site doped (K,Na)NbO_3 (KNBNNO), made in the author's laboratory. Photon energy of 0 eV represents the dark environment.

of conductivity under illumination compared to the value measured in the dark due to the already high concentration of mobile charge carriers. This case is well represented by a commercial p-type monocrystalline Si (purchased from Okmetic, Finland) shown in figure 5.4.

As for a classic ferroelectric material that can be considered both an insulator and a wide band gap semiconductor, e.g. a commercial PZT-5A (purchased from Piezo.com, USA), the conductivity in the dark can be as low as 10^{-11} S m^{-1}. Such a good insulator ensures the applied external electric field to generate negligible leakage current going through the sample and as a result, the electric field is able to effectively switch polarization and thus give a nicely measurable ferroelectricity, as shown in figure 5.5. On the other hand, incident light may generate mobile charge carriers in ferroelectrics. Since the spontaneous polarization bends the band structure in and around the domain walls, grain boundaries and other interfaces adjacent to the polarization, it creates a favoured route for charge carriers to migrate when subject to an external electric potential, as explained in figure 5.6. This has been known as the cumulative effect of light and electric field on the current modulation (Vats *et al* 2020). Consequently, the conductivity of the PZT-5A is able to increase by an order of magnitude under illumination (figure 5.4). The increase is much more obvious than that of the Si sample but it is only moderate compared to

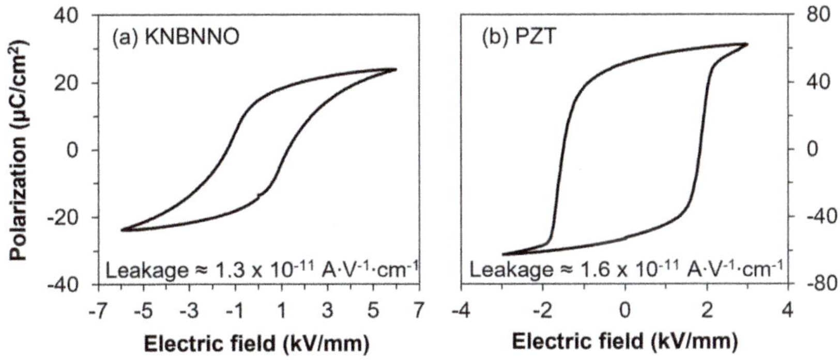

Figure 5.5. Ferroelectric hysteresis loops and leakage currents for (a) a KNBNNO ceramic sample made in laboratory and (b) a commercial PZT-5A ceramic sample.

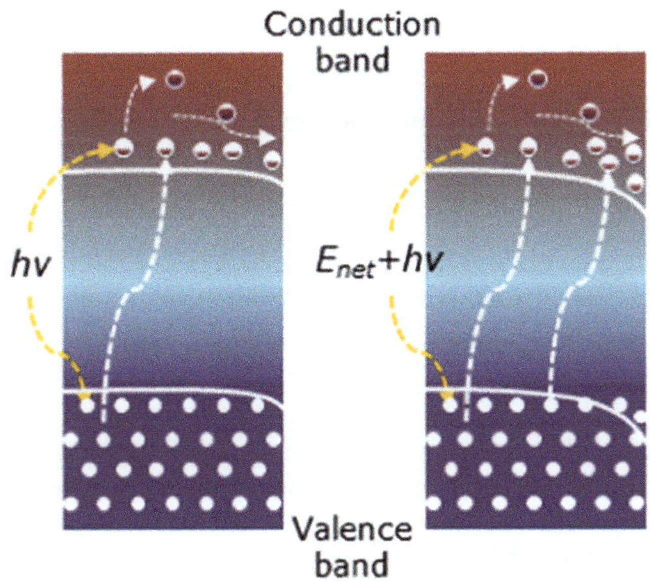

Figure 5.6. Schematics illustrating the mechanism of the optoelectrical cumulative effect on the conduction of photo-generated charge carriers. The right edge of the band structures is supposed to be a domain wall, a grain boundary or another interfacial defect. Reproduced under the CC-BY license from (Vats *et al* 2020).

narrow band gap ferroelectrics due to the PZT-5A sample's band gap of >3 eV (Bai *et al* 2017b, 2019a). All the given photon energies in figure 5.4 are below such a wide band gap and the potential photo-generated charge carriers are estimated to originate from absorption of in-gap states.

The light-electric field cumulative effect on the conductivity of ferroelectrics becomes tremendous when the band gap is reduced into the visible-light range. The Ba A-site Ni B-site doped (K,Na)NbO_3 (KNBNNO) as indicated in figure 5.4 exhibits a dark conductivity (roughly at the level of 10^{-6} S m^{-1}) between the values of the classic semiconductors and wide band gap ferroelectrics. The KNBNNO sample generally has a band gap of 1.6–2.2 eV depending on the stoichiometry (Bai *et al* 2017b, 2021). Reflecting equation (5.2), although the band gap is narrow, KNBNNO is still able to maintain a small enough dark conductivity to obtain an easily measurable ferroelectricity (figure 5.5), most likely owing to the low mobility of charge carriers. Indeed, the material retains the nature of an insulating ferroelectric based on the frame of oxide perovskite, given that the doping concentration is minimal (1–2 mol%). The simultaneous presence of a narrow band gap and low charge carrier mobility is crucial for the further development of BPVE and multi-energy harvesters/sensors, which will be discussed in the following sections.

However, because the photon energies of the incident light included in figure 5.4, especially 2.25 eV and 3.06 eV, are above the band gap of the material, a linear relationship between the KNBNNO's conductivity and incident photon energy can be seen for light energies above 1.88 eV. The increase of the conductivity under illumination is as high as three orders of magnitude, approaching the classic semiconductor's level.

Based on the above discussion, the modern definition of photoferroelectrics should include the phenomenon of conductivity being influenced by the optoelectrical cumulative effect. Despite the band gap value, a sufficiently large spontaneous polarization that is able to modify the band structures at the domain walls or other intergranular boundaries and thus modify the conductive path or the potential threshold to migrate charge carriers under applied electric field, is an essential factor.

5.2.2 Photo-excited and photo-stimulated domain wall motion

Figure 5.3 shows that ferroelectric domains can be re-oriented under illumination. This phenomenon is known as the photo-excited domain wall motion, or the optical control of domains (Rubio-Marcos *et al* 2018, Vats *et al* 2019). It has been observed for both above-gap and below-gap illuminations. For instance, the 532 nm (2.33 eV) incident light used in figure 5.3 has a photon energy smaller than the band gap of the $BaTiO_3$ single crystal (>3 eV). A similar behaviour is also observed for $BaTiO_3$ ceramics in which larger grains exhibited photostriction when exposed to the same 532 nm light (Rubio-Marcos *et al* 2021). The photostriction is believed to originate from photo-excited domain reorientation. The magnitude of photostriction is comparable to the material deformation induced by electric field, paving a promising way towards optical or optoelectrical dual-way actuators, i.e. opto-piezoelectric devices.

For above-gap illuminations, studies have been carried out on KNBNNO ceramics. Figure 5.7 shows piezoresponse force microscopy (PFM) images after mapping the surfaces of polished KNBNNO samples subject to illumination or electric field (Vats *et al* 2020). The photon energy of the incident light is 3.06 eV (405 nm wavelength), much higher than the material's band gap of 1.6–2.2 eV. The marked areas in each sub-figure clearly indicate the domain reorientation solely under illumination to the extent comparable to that driven solely by electric field, unambiguously proving photo-excited domain switching.

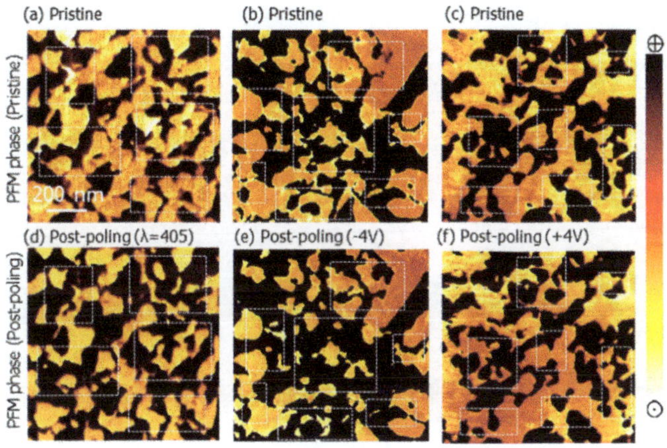

Figure 5.7. PFM images comparing optical and electrical responses of domains and domain walls. Out-of-plane phase images of (a)–(c) ferroelectric domains at three different locations under the dark condition, (d) post optical poling with a laser source of 405 nm wavelength with no electrical bias at the same location of (a), and post electrical poling in the dark with (e) −4 V (electric field −40 V mm^{-1}) and (f) +4V (electric field +40 V mm^{-1}) at the same locations of (b) and (c), respectively. Reproduced under the CC-BY license from (Vats *et al* 2020).

The optoelectrical cumulative effect on conductivity introduced in section 5.2.1 can also be reflected by a similar effect on ferroelectric switching. Figure 5.8 demonstrates that, surprisingly, a large area of domains that are exposed to light with an above-gap photon energy as well as being biased by a small electric field can manage to reorient, despite the fact that the electric field is only applied on a 30 nm-sized single point (note the sample surface is not covered by any electrode) (Vats *et al* 2020). According to figure 5.5, the electric field applied in figure 5.8 is only less than 3% of the coercive field. In other words, the incident photon energy makes up 97% of the necessary but missing electrical energy for ferroelectric switching.

A domino effect is also present. Light cannot individually reorient such a large number of domains based on the observation in figure 5.7, so that the switching in figure 5.8 must be due to the optoelectrical cumulative effect. Since the electric field is only applied to a single point, the domain(s) at this single point should reorient first. It is possible that such a move will cause inter-domain stress and with the help of the photon energy, the stress that normally would not be powerful enough to make any difference will now be able to trigger domain wall motion (domain reorientation) in the adjacent spaces. The subsequent domain wall motion will further trigger other domain wall reorientation events until the edge of the illumination spot where the supplied photon energy disappears.

The phenomenon demonstrated by figure 5.8 is referred to as the photo-stimulated (or light-assisted) domain wall motion. It appears that the incident photons themselves already have a preferred direction for assisting the domain wall motion because the cumulative effect with an electric field in the same direction to

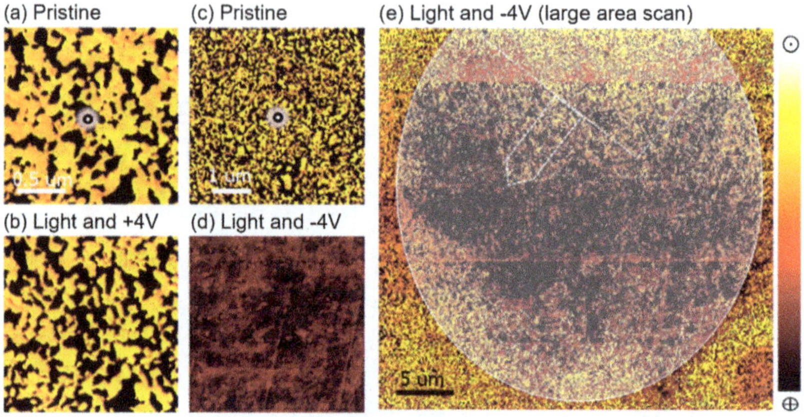

Figure 5.8. Cumulative effect of the light and applied electric field on ferroelectric switching. (a), (c) PFM phase images under the dark condition. Changes in the domain structure when subject to both incident light (405 nm laser) and external electric field of (b) +4 V (+40 V mm^{-1}) and (d) −4 V (−40 V mm^{-1}). The locations in (a) and (b), and in (d) and (c), respectively, are the same. The electric field is applied on a single point as marked in (a) and (c) with a 30 nm probe tip. (e) Large area scan under illumination of the 405 nm laser and −4 V electric field for an extended period (40 s). The white shaded area indicates illumination area and the white dashed lines shows the probe position. Reproduced under the CC-BY license from (Vats *et al* 2020).

that of the incident photon flux is considerably stronger than that with an electric field in the opposite direction.

In summary, the modern definition of photoferroelectrics should also involve photo-excited and photo-stimulated domain wall motion.

5.2.3 Bulk photovoltaic effect and ferroelectric photovoltaics

The third aspect of the modern definition of photoferroelectrics is undoubtedly the earliest known aspect of light–ferroelectric interaction, that is the BPVE.

BPVE has a fundamentally different working mechanism for converting photon energy into electricity compared to semiconductor solar cells which rely on a built-in potential (e.g. p–n junction, Schottky barrier) to separate charge carriers. For BPVE, the photo-excited charge carriers are generated with a built-in net momentum in ferroelectrics (or non-centrosymmetric materials). This momentum allows the charge carriers to travel the mean free path (l_0) before thermalizing to the band edge to create a directed photogalvanic current (Lopez-Varo *et al* 2016). Unlike semiconductor solar cells for which the mechanisms are clear and widely agreed upon, a convincing and universal theory for BPVE has not been established. The 'ballistic current mechanism' is a popular theory to describe the BPVE, which is used here to guide the discussion. The short-circuit photocurrent density (J_{SC}), open-circuit photovoltage (V_{OC}) and the PCE (η) can be expressed by equations (5.3)–(5.5) (Lopez-Varo *et al* 2016).

$$J_{SC} = q l_0 \xi \varphi \alpha \Phi \tag{5.3}$$

$$V_{OC} = \frac{l_0 \xi d}{(\mu_e + \mu_h)\tau} \tag{5.4}$$

$$\eta = \frac{q\alpha\varphi(l_0\xi)^2 d}{(\mu_e + \mu_h)\tau h\nu} \cdot FF \tag{5.5}$$

In these equations, q is the positive elementary charge, ξ is a measure of the excitation asymmetry, φ is the quantum yield for charge generation, α is the absorption coefficient, Φ is the photon flux density of the incident light, d is the distance between the two contacts (electrodes), τ is the charge carrier lifetime and FF is the fill factor. h is Planck's constant, ν is the wavelength of the incident light and thus $h\nu$ is the photon energy of the incident light.

It is estimated from equation (5.4) that a larger d induces a higher V_{OC}. Unlike the V_{OC} in semiconductor (junction-based) solar cells that is limited by the band gap of the absorber (Würfel and Würfel 2016), for BPVE nothing prevents the V_{OC} from exceeding the band gap if d is large enough (Lopez-Varo *et al* 2016, Paillard *et al* 2016, Spanier *et al* 2016). Figure 5.9 shows an example of the generation of V_{OC} exceeding the band gap (Seidel *et al* 2011). Two strip electrodes are made on the same surface of a 100 nm thick $BiFeO_3$ film separated by a distance (d value) of

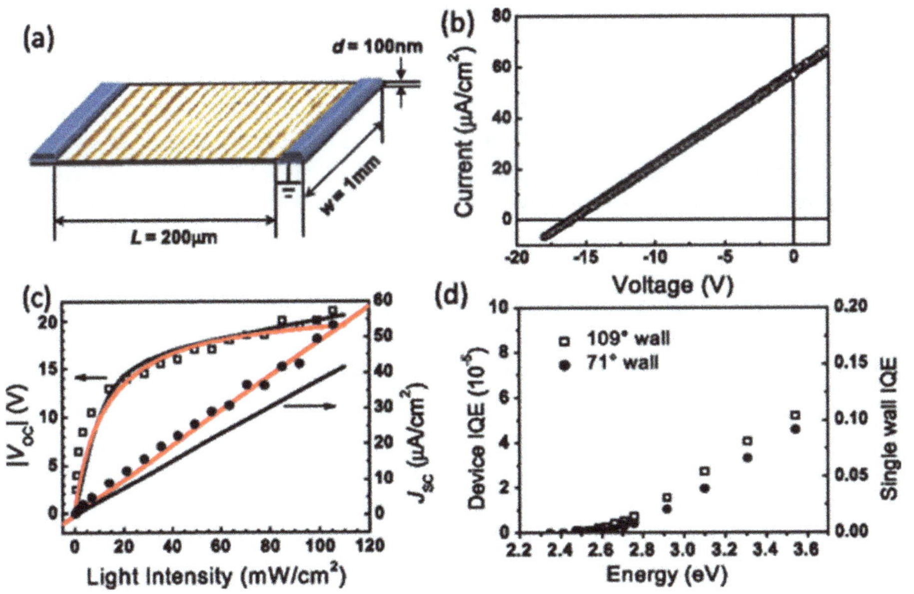

Figure 5.9. (a) Schematic of the $BiFeO_3$ film device configuration. (b) Measured J–V curve under white light illumination (200 mW cm^{-2}). (c) Measured V_{OC} and J_{SC} for 375 nm illumination with black solid lines for calculated results using the numerical model and the red lines for fits to the analytical model. (d) Measured IQE (internal quantum efficiency) for the device and for a single domain wall. Reproduced with permission from Seidel *et al* (2011). Copyright © 2011 American Physical Society.

200 μm (i.e. transverse structure). An ultrahigh V_{OC} of >15 V (75 V mm^{-1}) in the J (current density)–V (voltage) curve, more than five times greater than the band gap, is achieved under 200 mW cm^{-2} white light. The above-gap photovoltage is a unique feature and fundamental advantage of BPVE compared to semiconductor photovoltaics.

As for the transverse configuration (figure 5.9), the domains between the two electrodes possessing a large d value have to be well aligned to guarantee a net polarization in parallel with the photocurrent flow. Otherwise, the small l_0 value will lead to trapping of the charge carriers locally and no net photocurrent will be extracted (Yang *et al* 2010, Seidel *et al* 2011). However, according to equation (5.5), to achieve a large PCE the material should also have a good absorption of incident light. It is thus difficult to simultaneously satisfy both factors of (i) a material having a narrow or tunable band gap and (ii) the domains in the material being easily manipulated to form directed net polarization over a long distance between the electrodes.

An alternative is to make ferroelectric photovoltaics (not necessarily BPVE) by sandwiching a thin ferroelectric film of a thickness of 100 nm or less between a pair of planar electrodes (i.e. plate capacitor structure). This practice mainly targets obtaining a satisfactory l_0. Because the l_0 values of most ferroelectrics are suggested to be 10–100 nm, in this case, all photo-excited charge carriers are non-thermalized and thus contribute to the photocurrent. As a result, record η values of 3.3% and 2.5% have been reported for 100 nm thick Bi_2FeCrO_6 (BFCO, band gap 1.4 eV) films under AM 1.5G light and 12 nm thick $Pb(Zr_{0.2}Ti_{0.8})O_3$ (band gap 3.6 eV) films under UV light, respectively (Nechache *et al* 2015, Tan *et al* 2019). These high η values (among oxide perovskite ferroelectrics) are attributed to relatively high J_{SC} (1–10 mA cm^{-2}). However, the V_{OC} values are smaller than 1 V due to the low value of the parameter d according to equation (5.4). Due to the small thickness, it is difficult to determine whether the photocurrent is derived from the heterojunction formed between the films and electrodes or from the BPVE. One may assume that there is at least partial BPVE contribution because of the presence of spontaneous polarization.

However, there is an intrinsic problem in both of the above thin film configurations. Oxide perovskite ferroelectrics normally have small α values at the edge of band gaps and hence need much thicker films to completely absorb the relatively low photon energy part of the incident light (Xu *et al* 2009, Nechache *et al* 2015). This forces one to employ the tandem concept used by multi-junction solar cells, losing the advantage of simple device configuration theoretically enabled by the BPVE (i.e. only a ferroelectric plus two electrodes compared to the complex electrode–transporter–absorber–transporter–electrode p–n configuration for semiconductor photovoltaics). For instance, the PCE of the BFCO films can be boosted from 3.3% for a single layer to 8.1% for a triple-layer structure under the same light source where the improvement is due to a more complete light absorption achieved by the different band gaps of the individual layers (Nechache *et al* 2015).

Despite the early achievements of relatively high PCEs in the plate capacitor structure, a further increase of the film thickness beyond the l_0 in the plate capacitor structure is likely to rapidly degrade J_{SC} at a pace that cannot be counteracted by the only moderate increase of V_{OC} to compensate the PCE (Tan *et al* 2019). This prevents the ferroelectric films from becoming more efficient. Instead, in principle the transverse configuration is more promising for future BPVE-based devices to increase the PCE to a level comparable to semiconductor-based photovoltaics. Although the current η values of the transverse configuration are not comparable to those of the plate configuration (e.g. ~0.1% compared to 2%–4%), the former offers a fundamentally larger potential to engineer with tuneable band gaps. For instance, the average α value of the $BiFeO_3$ film for visible lights is about 10^3 cm^{-1}, requiring a 10 μm thick material to completely absorb the incident light (Xu *et al* 2009, Nechache *et al* 2015). This means in the $BiFeO_3$ film ($\eta \approx 0.1\%$) in figure 5.9, only 1% of the incident light intensity has been absorbed. Engineering the band gap and hence the α value while retaining the ferroelectric nature of the materials is the key to achieving high-efficiency BPVE devices, as will be discussed in the next section.

Another significant difference between the ideal/classic BPVE and the semiconductor photovoltaics is that the *FF* value of BPVE is likely to be 0.25 for the former (Lopez-Varo *et al* 2016). An example is that the *J–V* curve is a straight line as shown in figure 5.9(b). However, the low *FF* should not be an issue for achieving a high PCE of BPVE because the photovoltage is not limited and thus can make up the loss of output power caused by a small *FF* value.

In summary, the term 'photoferroelectrics' is defined as ferroelectric materials showing any of these phenomena: (1) drastic increase of conductivity under above-gap photon energy; (2) photo-excited and/or photo-stimulated domain wall motion; and (3) BPVE.

5.3 Band gap engineering of ferroelectrics

Engineering and thus reducing the currently wide band gaps of good ferroelectrics will considerably broaden the application of photoferroelectrics as the widely available visible light will interact with ferroelectricity. Although it is not difficult to narrow down the band gap of ABO_3 perovskites by doping, most band gap engineering methods reported so far come at the cost of compromising the ferroelectricity. It is known from the above discussion that a decent ferroelectricity is a prerequisite for photoferroelectrics. Therefore, band gap engineering of ferroelectrics must be carefully designed and studied to achieve an optimal trade-off of light absorption and photoferroelectric properties.

5.3.1 Why do good ferroelectrics have wide band gaps?

From chapter 2, the theoretical reasons for wide band gaps of ferroelectrics that possess strong spontaneous polarizations are known. The hypothesis has also been validated in experiments (Grinberg *et al* 2013, Bai *et al* 2017a, 2017b). Briefly speaking, most classic ferroelectric compounds which show clear and reliable switchability of the strong spontaneous polarizations belong to asymmetric space groups with a relatively large displacement of the B-site cation from the centre of the oxygen octahedron. Since d^0 transition metals are usually required on the B site to obtain a strong ferroelectricity, as is the case in widely studied compositions including $BaTiO_3$, $Pb(Zr,Ti)O_3$, $BiFeO_3$, $KNbO_3$ and $(K,Na)NbO_3$, the B–O bond has a large electronegativity difference which gives rise to large spontaneous polarization as well as extremely low conductivity (section 5.2.1). Such a B–O bond also imposes a high energy cost on the transition of the electrons from the valence band minimum (VBM) to the conduction band maximum (CBM), leading to a wide band gap. Of the classic ferroelectrics, the rhombohedral-phase $BiFeO_3$ has the smallest band gap (2.7 eV) and largest remanent or spontaneous polarization (~50 μC cm^{-2}) depending on the form of the material (e.g. ceramics, single crystals or thin films).

Both the ferroelectricity and band gap are determined by the same B–O bond in these materials. This fact seems to point to the inevitability of compromising ferroelectricity to achieve visible-light absorption. Nevertheless, there are still some experimentally proven approaches to avoid this fate, at least to some extent.

5.3.2 Band gap engineering strategies

To retain reasonable ferroelectric properties while reducing the band gaps, B-site ordering, B-site defect dipole and micro-heterojunction are the three practical routes to follow.

B-site ordering—BFCO

The double perovskite Bi_2FeCrO_6 (BFCO) is derived from the parent $BiFeO_3$ by replacing half of the B-site Fe^{3+} cations with the same number of Cr^{3+} cations. The ordering of Fe^{3+} and Cr^{3+} on the B site plays a crucial role in tuning the band gap. While the ferroelectricity of BFCO is driven mostly by the Bi^{3+} A-site cations, the band gap is controlled by the interaction between Fe^{3+} and Cr^{3+} via O^{2-} which alternate on the B site. The key here is to make the mechanisms of ferroelectricity and band gap reduction occur simultaneously at different sites. This brings special requirements for the fabrication of the BFCO materials. Figure 5.10 illustrates the dependence of the photoresponse on the Fe/Cr ordering (Nechache *et al* 2015). With different Fe/Cr orderings, the BFCO films exhibit distinct optical behaviours with

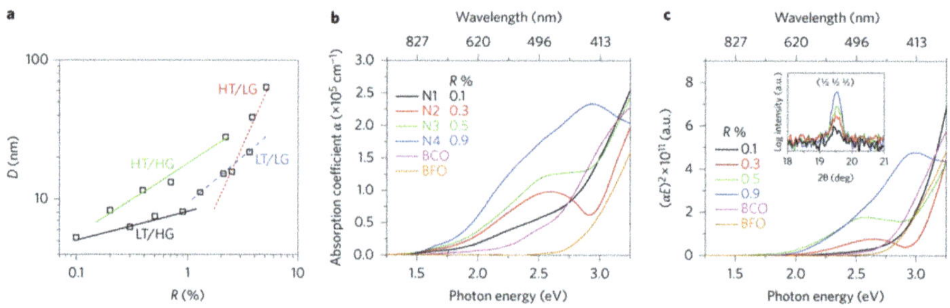

Figure 5.10. (a) Relationship between the R ratio (Fe/Cr cation ordering) and D (film thickness) in the BFCO film growth process using pulsed laser deposition. HT—high temperature; LT—low temperature; HG—high growth rate; LG—low growth rate. (b) Corresponding absorption coefficients and (c) direct optical transitions for the BFCO films grown with different R ratios. Reproduced from Nechache *et al* (2015). Copyright © 2014 with permission of Springer Nature.

band gaps ranging from 1.4 eV to >2 eV compared to ~2.7 eV for both the pure $BiFeO_3$ (BFO) and $BiCrO_3$ (BCO). The remanent polarizations of the BFCO films are measured to be 10–40 μC cm^{-2} with up to 80% retention of polarization compared to BFO.

The strict requirement for the Fe/Cr ordering is already difficult to handle during the growth of thin films and will be even more challenging for the fabrication of bulk materials such as single crystals and ceramics. Therefore, further exploitation of the BFCO as a photoferroelectric heavily relies on in-depth studies of methods for reliable fabrication of this material.

B-site defect dipole—Ni^{2+}-oxygen vacancy

It has been predicted by DFT calculations (Grinberg *et al* 2013) that through replacing Nb^{5+} in $KNbO_3$ with Ni^{2+} and paring the Ni^{2+} with an oxygen vacancy (V_O), the direct band–band transition can be shifted from the O 2p-Nb 4d to the Ni 3d-Nb 4d. This approach indeed works, however, researchers may encounter problems of either only moderately reduced band gap/no band gap tuning at all, or complete disappearance of ferroelectricity (Bai *et al* 2019a, Lou *et al* 2019, Wu 2019, Xiao *et al* 2019, Zhong *et al* 2019). Researchers need to pay attention to several factors simultaneously in order to achieve ideal results:

(1) This method is highly sensitive to composition (Bai *et al* 2017a, 2017b, Wu *et al* 2019). One cannot assume that the Ni^{2+}–V_O doping can be used in any oxide perovskite composition to achieve the same outcome. While some other compositions have been tried, the band gap reduction effect is doubtful. Detailed discussion will be given in section 5.3.3 regarding possible counterfeit band gaps.

(2) The doping should be precise. Similar to the strict requirement for Fe/Cr ordering in BFCO, ideally exactly the same amounts of Ni^{2+} and V_O should be generated to be paired with each other (Bai *et al* 2017b, 2021). Any excessive Ni^{2+} or V_O is very likely to significantly degrade the ferroelectricity as well as ruin the band gap reduction. This is challenging for solid-state reactions via which most oxide perovskite ferroelectric ceramics are made.

(3) A critical value for the doping concentration should be found. Because in theory, the Ni^{2+} is directly replacing Nb^{5+} ($KNbO_3$ or $(K,Na)NbO_3$ as an example) which tends to weaken the second-order Jahn–Teller effect of the B–O bond, in addition to the fact that V_O is likely to induce domain wall pinning effect, the doping concentration should be precisely controlled to be neither too small in order to provide enough energy states for band gap reduction nor too large in order to retain ferroelectricity (Bai *et al* 2017b, 2019b). This, again, leads to challenges with regard to fabrication for the solid-state route.

(4) In practice, the V_O is naturally formed by co-doping a proper concentration of Ba^{2+} on the A site of $KNbO_3$ or $(K,Na)NbO_3$, and then sintering the ceramic samples in air without special atmospheric control.

When carried out correctly, the method of introducing Ni^{2+}–V_O defect dipoles into the $KNbO_3$ or $(K,Na)NbO_3$ unit cells is able to narrow down the band gap by up to ~2 eV while retaining roughly 50% of the parent oxide polarization (Bai *et al* 2017b).

Micro-heterojunction

Formation of a heterojunction is widely used in semiconductors to tune the band structures at the junction interface. Instead of contacting two monolithic semiconductor materials at the macroscale, heterojunctions can also be employed in photoferroelectrics but at the microscale, i.e. to be formed at domain walls and grain boundaries where a large number of micro-heterojunctions can be formed inside the material. Figure 5.11 visualizes an early concept of micro-heterojunction formed between $BiMnO_3$ and $BiMn_2O_5$ in a Bi–Mn–O composite (Chakrabartty *et al* 2018). The $BiMnO_3$ is a ferroelectric phase and the $BiMn_2O_5$ is a semiconductor phase with a band bap of ~1.2 eV. The interface between the two phases results in the coexistence of ferroelectricity and narrow band gap.

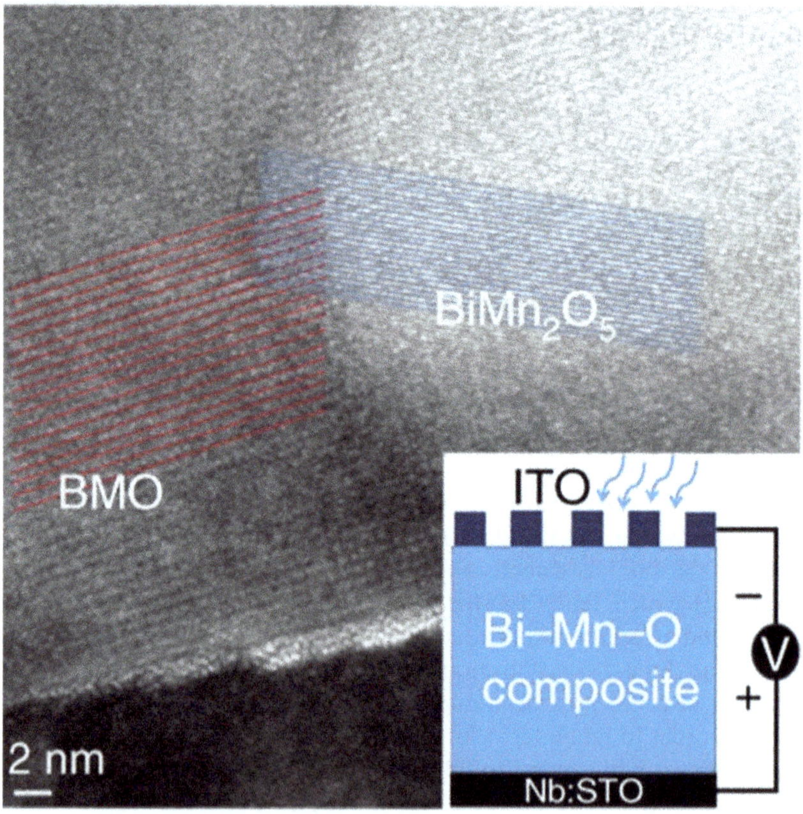

Figure 5.11. TEM (transmission electron microscopy) image showing the simultaneous presence of BMO ($BiMnO_3$) and $BiMn_2O_5$ crystal phases in a Bi–Mn–O composite thin film. The inset shows a schematic of the device (cross-sectional view) and the measurement configuration where ITO is indium tin oxide used for the top transparent electrode, STO is strontium titanate and Nb:STO is used as the conductive substrate on which the film is grown. Reproduced from Chakrabartty *et al* (2018). Copyright © 2018 with permission of Springer Nature.

Apart from thin films, the micro-heterojunction has been further developed and has been found to be especially helpful in photoferroelectric ceramics where precise control of atomic ordering or doping is complicated, as discussed above. Figures 5.12–5.15 summarize the systematic research on the micro-heterojunction which is in fact a pseudo-morphotropic phase boundary (MPB) formed between the perovskite phase and the tungsten bronze phase in KNBNNO ceramics (Bai *et al* 2021). In practice, the success rate of achieving replicable, precise atomic control required by the B-site ordering and defect dipole approaches for band gap engineering is relatively low. However, replicating certain type of composition within a tolerance window, as shown in figure 5.12, can be easily achieved. The designed compositions are based on the frame of $(K_i,Na_j,Ba_k)(Nb_l,Ni_m)O_{3-\delta}$ in which the values of k, m and δ are no longer precisely required, instead, variation within the window of each type (A, B, C, D) is acceptable.

Figure 5.13 displays the resultant four types of ceramic samples. The optical properties of different types of samples are already distinct even without the help of absorption measurements. Type A has a pure perovskite KNBNNO phase while

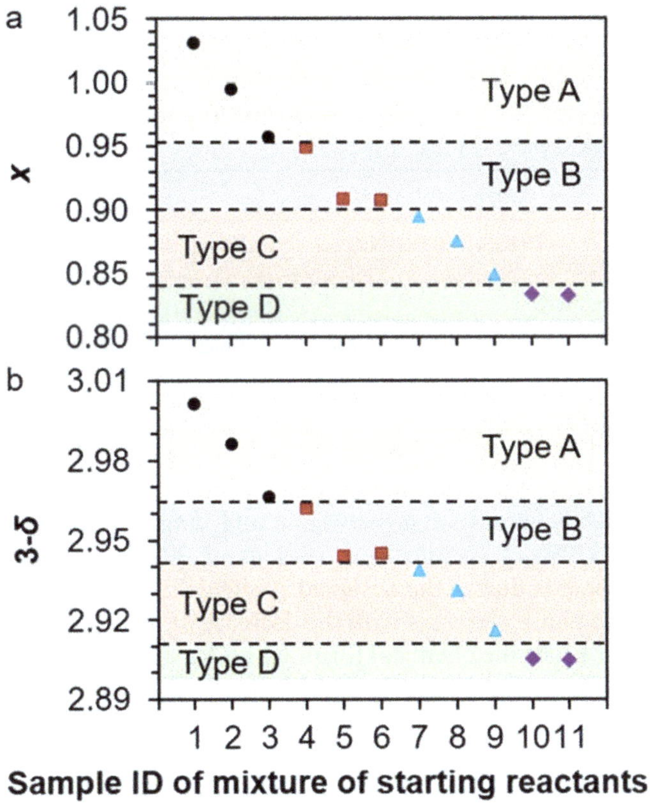

Figure 5.12. Different values of x (A-site molecular number in the $ABO_{3-\delta}$ normalized composition) and $3-\delta$ (oxygen molecular number, δ represents the molecular number of V_O) inducing four types of $(K,Na,Ba)_x(Ni,Nb)O_{3-\delta}$ ceramics. Reproduced under CC-BY license from (Bai *et al* 2021).

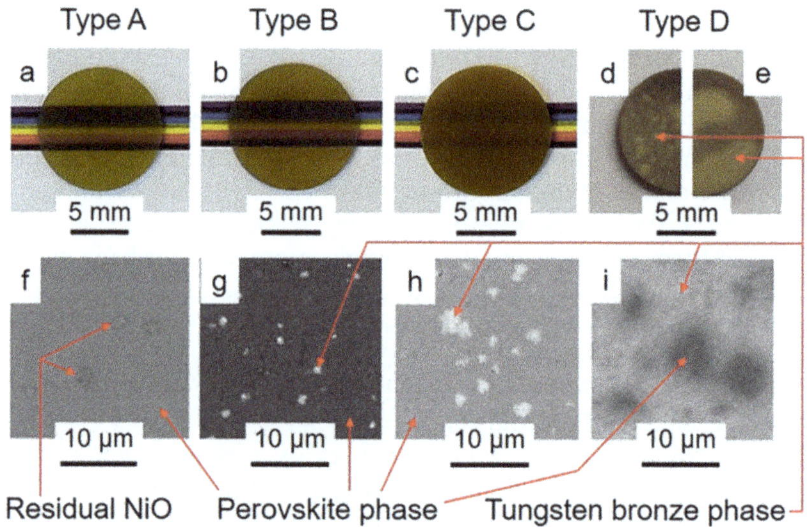

Figure 5.13. (a)–(e) Photographs and (f)–(i) backscattered scanning electron microscopy (SEM) images of (a), (f) Type A, (b), (g) Type B, (c), (h) Type C and (d), (e) and (i) Type D ceramics corresponding to those classified in figure 5.12. Reproduced under CC-BY license from (Bai *et al* 2021).

Table 5.1. Chemical formulas and space groups of the KNBNNO perovskite and tungsten bronze phases. Reproduced under CC-BY license from (Bai *et al* 2021).

Phase	Chemical formula	Space group
Perovskite	$(K_{0.537 \pm 0.023}Na_{0.374 \pm 0.021}Ba_{0.028 \pm 0.003})$ $(Nb_{0.996 \pm 0.001}Ni_{0.004 \pm 0.001})O_{2.977 \pm 0.007}$	Amm2 Orthorhombic
Tungsten bronze	$(K_{0.251 \pm 0.021}Na_{0.097 \pm 0.013}Ba_{0.190 \pm 0.014})$ $(Nb_{0.970 \pm 0.003}Ni_{0.030 \pm 0.003})O_{2.818 \pm 0.011}$	P4bm Tetragonal

Types B, C and D have co-existing perovskite and tungsten bronze phases, both of which are the KNBNNO composition (Bai *et al* 2021). The formation of the tungsten bronze phase is due to the designed A-site deficiency. The concentration of the tungsten bronze phase increases with the decrease of the x value, i.e. from Type B to Type D. Table 5.1 lists the chemical formulas and space groups of the two phases.

The interfaces present between the two phases form micro-heterojunctions that are randomly and evenly distributed inside the ceramic samples. Owing to these micro-heterojunctions, the band gaps of Type B and Type C samples can be reduced by 1–2 eV compared to those of Type A and Type D samples, as illustrated in figure 5.14. While the band gap reaches the minimum value, the polarization jumps to the optimum level, as shown in figure 5.15. This trend is different compared to the B-site defect dipole approach where the band gap and polarization usually both

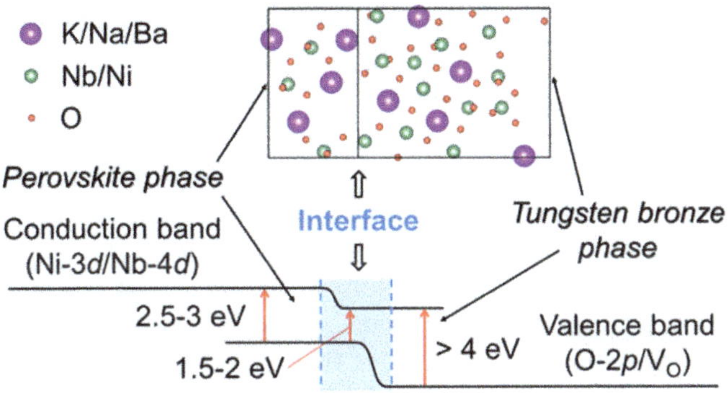

Figure 5.14. Schematics of the interface between the perovskite and tungsten bronze phases with the micro-heterojunction narrowing down the band gap of either individual phase. Reproduced under CC-BY license from (Bai *et al* 2021).

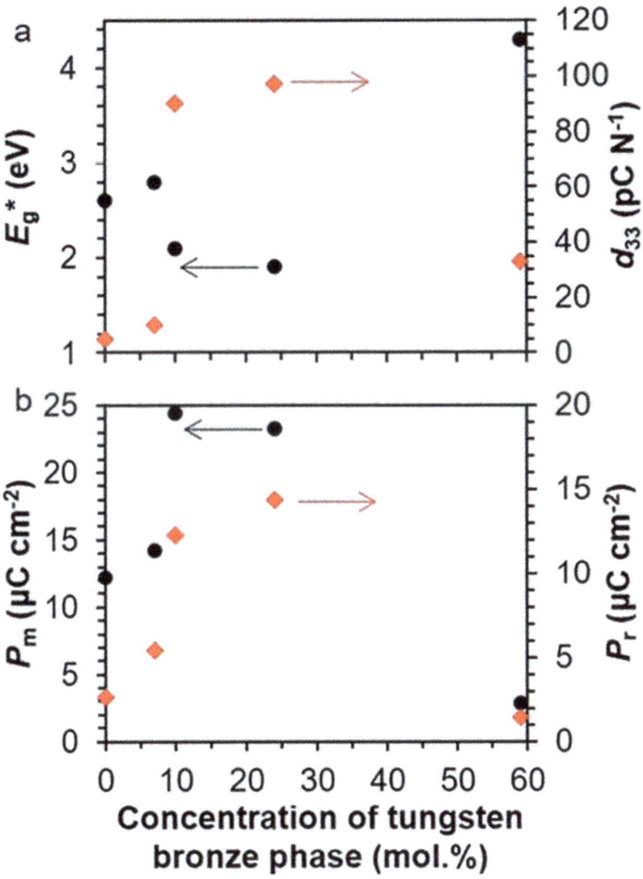

Figure 5.15. Dependence of (a) effective band gap (E_g^*) and piezoelectric coefficient (d_{33}) and (b) polarization at the maximum applied electric field (P_m) and remanent polarization (P_r) on the concentration of the tungsten bronze phase. Reproduced under CC-BY license from (Bai *et al* 2021).

decrease which is unfavoured in applications, even though both approaches are based on the KNBNNO compositions but with different exact chemical formulas. The optimized polarizations in figure 5.15 also benefit from the micro-heterojunctions where the polar perovskite phase and the polar tungsten bronze phase construct a pseudo-MPB which is different from a true MPB formed between two ferroelectric perovskite phases but works through a similar mechanism by relaxing the energy barrier for domain reorientation (Bai *et al* 2021). The retained polarization is approximately 75% of the pristine value of the parent $(K,Na)NbO_3$.

Figure 5.15 also provides evidence that the band gap engineering while retaining the ferroelectricity is achieved by neither of the individual phases but rather is due to the interfaces, as either a sufficiently high or low concentration of the tungsten bronze phase leads to degraded results for band gap reduction and polarization.

In summary, all three strategies discussed above are similarly effective for experimental band gap engineering while retaining ferroelectricity with the third approach being more practically viable for making ceramics via the cost-effective and industrially used solid-state reaction.

5.3.3 Counterfeit band gaps in ferroelectric ceramics

Optical measurement of absorbance and/or reflectance using the Tauc plot has been widely used as a reliable characterization tool to define the band gaps of classic semiconductors. However, such optical spectroscopy is no longer reliable for the band gap determination of photoferroelectrics due to the influence of grain boundaries, domains and pores. Figures 5.16–5.18 showcase an example of microstructural pores with sizes comparable to the wavelengths of incident light distorting the results of optical spectroscopy and consequently misleading the band gap identification (Bai 2021a). In this example, four different KNBNNO ceramics samples fabricated under different conditions are compared: Type I sintered at 1130 °C and 1140 °C, and Type II sintered at 1150 °C sealed and in air atmosphere, respectively. Figure 5.16 clearly reveals porous and dense microstructure for Type I samples sintered at 1130 °C and 1140 °C, respectively. It also shows dense grains and the grains with inner-granular pores in Type II samples sintered under sealed condition and in air, respectively.

According to figure 5.17, Type I-1130 °C should have better absorption than Type I-1140 °C and band gaps appear to exist at approximately 1.6 eV and 2.5 eV for the former compared to 3 eV for the latter. Similarly, Type II-air shows better absorption than Type II-sealed and the former seems to have band gaps of 1.6 eV and 2.1 eV compared to the 2.2 eV of the latter. In fact, these conclusions are not valid. The additional absorption peaks seen for the porous Type I-1130 °C and Type II-air samples are caused by the interference of the pores with incident light for which the wavelength is comparable to the pores' sizes (Bai 2021a).

Conductivity measurement can obtain a more valid estimate of band gaps compared to optical spectroscopy. As shown in figure 5.18, the conductivity under illumination does not even increase by an order of magnitude from the dark value for photon energies below 2.6–2.9 eV for both the Type I samples. The additional

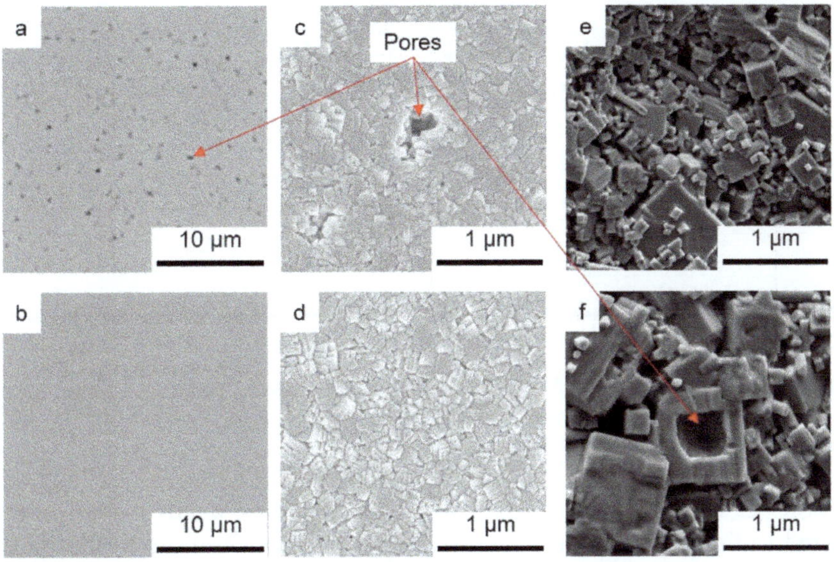

Figure 5.16. FESEM (field emissions scanning electron microscopy) micrographs of (a), (c) Type I KNBNNO ceramics sintered at 1130 °C, (b), (d) Type I KNBNNO ceramics sintered at 1140 °C, (e) Type II ceramics sintered at 1150 °C under sealed condition, and (f) Type II ceramics sintered at 1150 °C in air atmosphere. Reproduced under CC-BY license from (Bai 2021a).

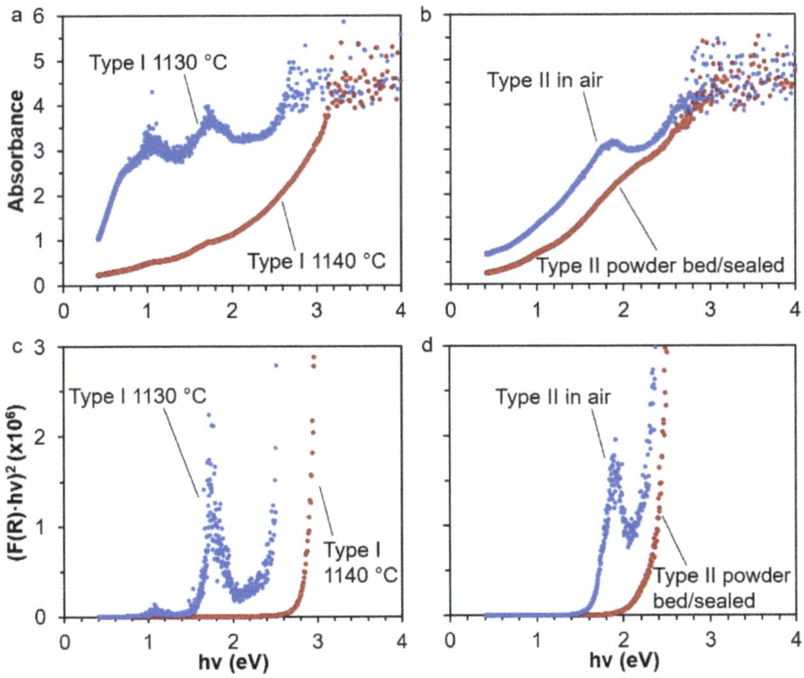

Figure 5.17. Dependence of (a), (b) absorbance on incident photon energy and (c), (d) the Tauc plots for (a), (c) Type I and (b), (d) Type II samples. R is reflectance; $F(R) = (1-R)^2/2R$; hv is photon energy of incident light. Reproduced under CC-BY license from (Bai 2021a).

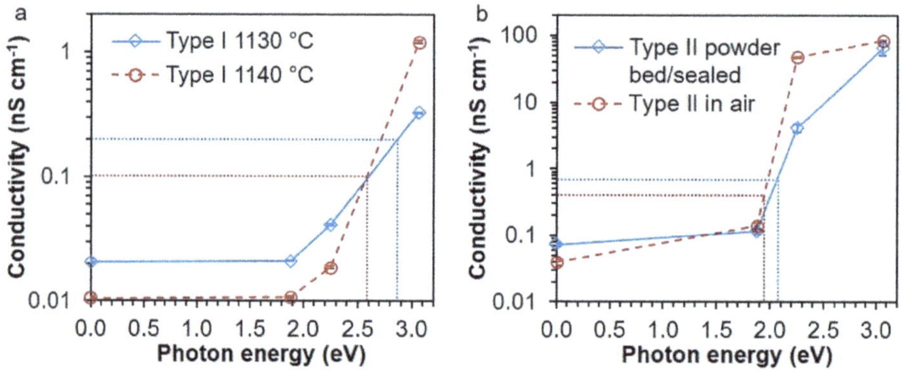

Figure 5.18. Dependence of conductivity on incident photon energy for (a) Type I and (b) Type II samples. Reproduced under CC-BY license from (Bai 2021a).

absorption peaks are not observed at the corresponding incident photon energies. Compared to figure 5.4, Type I samples behave rather similarly to the PZT which has a known band gap of 3 eV and no major difference has been found between the two Type I samples regardless of the differing optical behaviours. Therefore, there is a risk of mis-identifying counterfeit band gaps which may underestimate the true band gaps by relying solely on the optical method. The band gaps of both the Type I samples are still considered to be wide, but the same logic applies to Type II samples. By matching the information provided in figure 5.4 one can estimate the band gaps of both the Type II samples to be close to 2 eV with no sign of additional absorption for the Type II-air samples.

Based on the lessons learnt above, researchers making photoferroelectric ceramics should first try everything to obtain dense samples free from microstructural voids to avoid the possible resultant disturbance in optical spectroscopy. More importantly, conductivity measurement under different illumination conditions should always be carried out along with the optical measurements to help to define the true band gaps. Band gap values only derived from the Tauc plot should not be fully trusted in the case of photoferroelectrics.

5.4 Multi-source energy harvesting/sensing based on photoferroelectrics

With the rapid development of WSN (wireless sensor networks) and IoT (Internet of Things), the number of sensors in each device, system and the entire society will increase exponentially. It is estimated that trillions of sensors will be installed in the next decades (Kanoun 2018). Changing batteries for these sensors will become impractical and unpredictably costly in terms of labour resources (Enocean 2015). Therefore, the lifespan of batteries should be significantly extended. Ideally, the role of batteries as the power supply of smart sensing systems should be taken over by other technologies. Energy harvesting is considered a proper substitution candidate and even the ultimate solution for the issue of battery lifespan (Bai *et al* 2018). Energy harvesters collect ambient energy (which otherwise will be wasted, such as light, kinetic or thermal energy) from the working environment of the WSN, and then convert it into electricity to provide long-term (typically >20 years) power sources.

Piezoelectric energy harvesters/nanogenerators made from piezoelectric materials to harvest kinetic energy, and solar cells/light energy harvesters made from photovoltaics to harvest solar or artificial light energy are some of the popular energy harvesters. Pyroelectric materials and devices may be used to harvest thermal energy. Although they are yet to become as powerful as their piezoelectric or photovoltaic counterparts, pyroelectric sensors have been widely used in thermal and motion sensors for a long time. The corresponding pyroelectric energy harvesters can be a helpful addition to the mainstream harvester energy collection methods.

5.4.1 The unique opportunity to unify piezoelectric, pyroelectric and photovoltaic energy harvesters/sensors in single ferroelectric materials

Functional materials have always formed the core of electronics. To date, most components have been mono-functional because of the functional materials used. For instance, a physical sensor can usually be sensitive to only one stimulus, such as a mechanical sensor, thermal sensor or optical detector. The materials used in these sensors may be piezoelectric, pyroelectric or photovoltaic, respectively, and convert mechanical, thermal or optical energy (analog signal) into electric energy (digital signal). Although active sensors based on changes in the resistance (resistive sensors) can be multi-functional, they consume much more energy than the above-mentioned passive counterparts as they always require input voltage or current. The future 5G/6G-enabled IoT technologies require long-term, uninterrupted operation of smart sensors, challenging the current batteries' lifespan (only 4–8 $\mu W\ cm^{-3}$ long-term capacity compared to the $mW\ cm^{-3}$-level requirement). For energy-efficient design, sensors should be as passive as possible to reduce energy consumption. It is even desirable for the sensors to be able to generate electricity themselves by harvesting waste energy from their working environment, i.e. incorporating the energy harvesting technology. Due to the exploding number of smart sensors (from billions to trillions) expected in the near future, device miniaturization has become vital. Otherwise, the electronic systems would become too cumbersome and will generate large amounts of electronic waste detrimental to the environment if not properly recycled (which is already the case nowadays).

It is logical to predict that the famous Moore's Law—the number of transistors on a microchip doubles every two years—will eventually become invalid due to the incredibly large number of transistors in a limited space. This is also true for other functional components. These issues can be ultimately solved by constructing a multi-task device that only takes the space of a mono-functional, individual component, for instance, sensing multiple stimuli and harvesting multiple energy sources simultaneously (Bai *et al* 2018, 2019a, Bai 2021b). The performance of each individual function should not be compromised because of the multi-functionality. No sensor or energy harvester available today meets these requirements due to material limitations. Assembly of individual components is not the ultimate solution, because again, this will strongly increase the device size, weight and complexity.

Photoferroelectrics provide a unique opportunity to build a multi-functional sensor or energy harvester made from a single material or based on a single component. As has been introduced in chapter 1, poled ferroelectrics are piezoelectric and pyroelectric. This chapter has also discussed the BPVE arising from ferroelectrics. The multi-functionality of ferroelectrics, or photoferroelectrics to be exact, especially for the band-gap-engineered photoferroelectrics, naturally stands out among the other functional material families.

It is worth noting that this opportunity is considered to be unique based on several theoretical grounds. In principle, other material families may compete with photoferroelectrics as multi-functional sensors or harvesters but in practice they cannot. For instance, to unify good piezoelectric and photovoltaic effects, one might:

(1) Apply flexoelectric effect in the centrosymmetric Si and GaAs which are excellent light absorbers and solar cell materials. The flexoelectric effect can break the structural symmetry by introducing strain at nano-scale and thus create piezoelectric behaviour (Nguyen *et al* 2013, Yudin and Tagantsev 2013). However, the piezoelectric response is expected to be orders of magnitude smaller compared to that of ABO_3 ferroelectric perovskites (Zubko *et al* 2013).

(2) Use some other compositions such as ZnO, GaN, AlN and CdS, which are indeed simultaneously piezoelectric and semiconducting (Bowen *et al* 2014). However, they neither exhibit a comparable piezoelectric coefficient to that of ABO_3 piezoelectrics nor a visible-range band gap comparable to that of Si/GaAs compositions (Bai *et al* 2018).

(3) Develop piezoelectricity in narrow band gap organometal halide perovskites. Due to their high charge carrier mobility, organometal halide perovskite solar cells are able to show record PCEs of >20% (single-junction) (Green *et al* 2021). This is competitive to those of Si and GaAs based solar cells. To date, several organometal halide perovskites have been studied both theoretically and experimentally for piezoelectricity. The d_{33} values are estimated to range from 4.1 pC N^{-1} up to 248 pC N^{-1} with unit cells in a specific polar configuration (Liu *et al* 2016). However, only $CH_3NH_3PbI_3$ (band gap ~1.6 eV) has been proved to have robust ferroelectric and piezoelectric behaviours, with a measured piezoelectric energy harvesting figure of merit ($d_{33} \cdot g_{33}$, g_{33} is a piezoelectric voltage coefficient) of $\ll 1 \times 10^{-12}$ m^2 N^{-1} (Rakita *et al* 2017, Jella *et al* 2019, Pandey *et al* 2019, Röhm *et al* 2020) which is negligible compared to those of ABO_3 perovskites (up to 160×10^{-12} m^2 N^{-1} for single crystals and up to 20×10^{-12} m^2 N^{-1} for ceramics) (Li *et al* 2018, 2019). In addition, their low resistivity and the resultant high leakage makes it difficult to fully pole these materials and reliably characterize ferroelectric and piezoelectric properties (Rakita *et al* 2017, Shahrokhi *et al* 2020).

In contrast, the band gap engineering of photoferroelectrics, although inevitably degrading polarization at least to a minor extent, does not necessarily compromise the piezoelectricity as has been proven experimentally (Bai *et al* 2017b, 2021).

Meanwhile, looking back to equation (5.5), the PCE of the BPVE effect is actually favoured by lower dark conductivity. This contradicts the semiconductor photovoltaic theory in which higher charge carrier mobility and lifetime are favoured in order to achieve high PCE (Würfel and Würfel 2016). Theory of piezoelectricity also requires low conductivity and thus good electrical insulation for effective poling procedure as well as for ensuring surface charge migration in external circuits upon mechanical excitation rather than a large extent of charge throughput. Therefore, the unification of good piezoelectric and photovoltaic effects could not be realized before the investigation of photoferroelectrics.

5.4.2 Integrated energy harvesters/sensors

Obtaining a photoferroelectric material that co-exhibits good piezoelectric properties, pyroelectric properties and high-efficiency BPVE will open the door to truly integrated energy harvesters/sensors based on the same material platform. A general strategy to design such a material is as follows:

(1) Choose a parent ABO_3 perovskite composition showing both excellent piezoelectricity and pyroelectricity that is also suitable for band gap engineering (as discussed in section 5.3) from the database.
(2) Engineer the band gap to enable visible-light absorption.
(3) Optimize the device structures and configurations to maximize the piezoelectric figure of merit, pyroelectric figure of merit, and PCE.

To date, there has hardly been a successful example of this strategy mainly due to the need for further strong effort for boosting the PCE of BPVE. Figures 5.19 and 5.20 demonstrate a very early model of integrated multi-energy harvester/sensor based on a single photoferroelectric material. A KNBNNO ceramic patch which exhibits a $d_{33}{\bullet}g_{33}$ of ~1.2×10^{-12} m^2 N^{-1}, a pyroelectric coefficient of ~128 μC m^{-2} K^{-1} and a band gap of ~1.6 eV is embedded on a stainless-steel substrate with a tip mass to construct a piezoelectric cantilever widely used as an energy harvesting configuration (figure 5.19) (Bai *et al* 2020). The only difference is that the

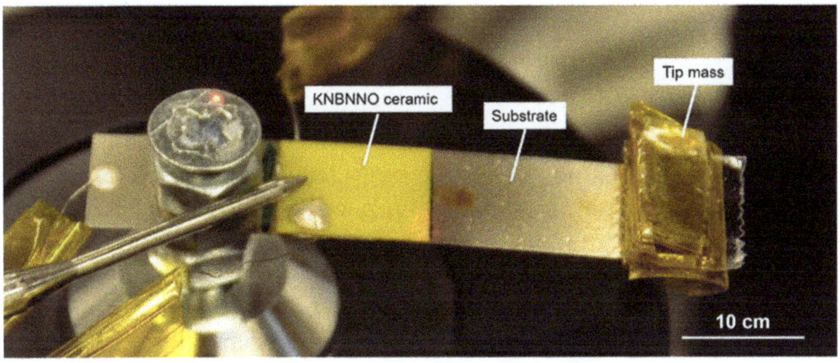

Figure 5.19. Image of an early model photoferroelectric multi-source energy harvester/multi-functional sensor. Reproduced with permission from Bai *et al* (2020). Copyright © 2020 John Wiley & Sons.

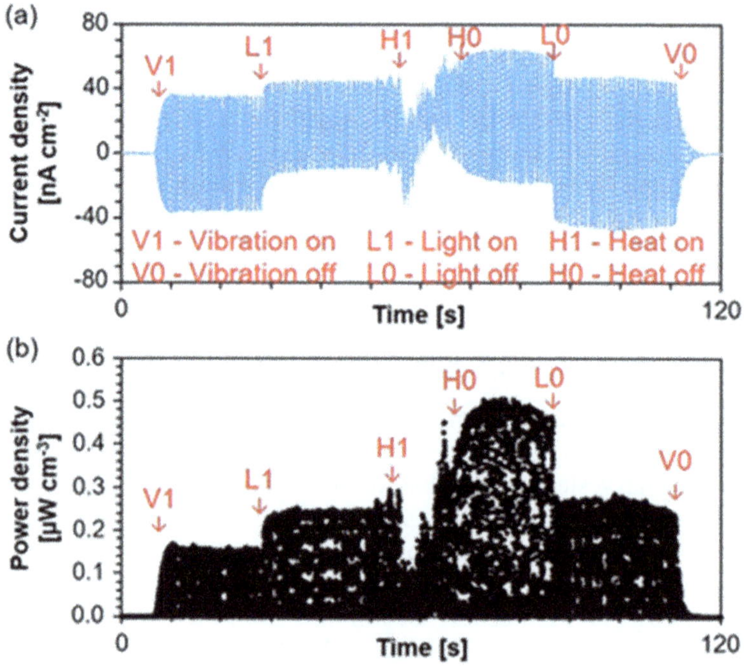

Figure 5.20. (a) Output current density and (b) output power density of the device shown in figure 5.19 with vibration, light, and heat input energy sources. Reproduced with permission from Bai *et al* (2020). Copyright © 2020 John Wiley & Sons.

conventional piezoelectric component is replaced by the KNBNNO photoferroelectric ceramic. During test, the devices are subject to harmonic vibration, temperature fluctuation and constant incident light (405 nm laser).

From figure 5.20, it can be clearly seen that the device is indeed able to harvest the three types of energy simultaneously and additively accumulate the output power from each source. However, as mentioned, this is only an early version and is nowhere near an ideal model. This leaves plenty of space for follow-up research and further development of photoferroelectric multi-energy harvesters/sensors.

5.5 Summary and perspectives

This chapter has discussed light–ferroelectric interaction and the modern definition of photoferroelectrics. A material can be considered to be photoferroelectric if it shows optical manipulation of conductivity, photo-excited/stimulated domain wall motion, and/or BPVE. Three practical band gap engineering strategies for photoferroelectrics have been described: through the B-site ordering suitable for thin films, the B-site defect dipole that is effective if the stoichiometry can be precisely controlled, and the micro-heterojunction showing excellent tolerance of off-stoichiometry and tunability of band gap/ferroelectricity/piezoelectricity. Band gaps identified solely through optical spectroscopy are highly susceptible to misjudgement by counterfeit signals, so that electrical conductivity should always be measured to assist the band gap identification.

In terms of applications, photoferroelectrics are unique candidates for integrated multi-energy harvesting/sensing. Further research should be carried out in the future to boost the BPVE efficiency as well as the capability of operation of the multi-functional device under different stimuli. For instance, how to add together or deconvolute the multiple electrical signals generated from the same material should become an essential topic for device engineering.

References

Bai Y 2021a Counterfeit band gaps caused by microstructural voids in photo-ferroelectric ceramics *Open Ceram.* **5** 100079

Bai Y 2021b Photoresponsive piezoelectrics *Front. Mater.* **8** 39

Bai Y, Jantunen H and Juuti J 2018 Energy harvesting research: the road from single source to multisource *Adv. Mater.* **30** 1707271

Bai Y, Jantunen H and Juuti J 2019a Ferroelectric oxides for solar energy conversion, multi-source energy harvesting/sensing, and opto-ferroelectric applications *ChemSusChem* **12** 2540–9

Bai Y, Kistanov A A, Cao W and Juuti J 2021 Vacancy-induced niobate perovskite-tungsten bronze composite for synergetic tuning of ferroelectricity and band gaps *J. Phys. Chem.* C **125** 8890–8

Bai Y, Palosaari J, Tofel P and Juuti J 2020 A single-material multi-source energy harvester, multifunctional sensor, and integrated harvester–sensor system—demonstration of concept *Energy Technol.* **8** 2000461

Bai Y, Siponkoski T, Peräntie J, Jantunen H and Juuti J 2017a Ferroelectric, pyroelectric, and piezoelectric properties of a photovoltaic perovskite oxide *Appl. Phys. Lett.* **110** 063903

Bai Y, Tofel P, Palosaari J, Jantunen H and Juuti J 2017b A game changer: a multifunctional perovskite exhibiting giant ferroelectricity and narrow bandgap with potential application in a truly monolithic multienergy harvester or sensor *Adv. Mater.* **29** 1700767

Bai Y, Xiang H, Jantunen H and Juuti J 2019b Multi-functional perovskites—an investigation of compositional and processing influence on microstructure, dielectric and ferroelectric properties *Eur. Phys. J. Spec. Top.* **228** 1555–73

Bhatnagar A, Chaudhuri A R, Kim Y H, Hesse D and Alexe M 2013 Role of domain walls in the abnormal photovoltaic effect in $BiFeO^3$ *Nat. Commun.* **4** 2835

Bowen C R, Kim H A, Weaver P M and Dunn S 2014 Piezoelectric and ferroelectric materials and structures for energy harvesting applications *Energy Environ. Sci.* **7** 25–44

Chakrabartty J, Harnagea C, Celikin M, Rosei F and Nechache R 2018 Improved photovoltaic performance from inorganic perovskite oxide thin films with mixed crystal phases *Nat. Photonics* **12** 271–6

Chanussot G and Glass A M 1976 A bulk photovoltaic effect due to electron-phonon coupling in polar crystals *Phys. Lett.* A **59** 405–7

Cheng S *et al* 2020 Highly controllable and silicon-compatible ferroelectric photovoltaic synapses for neuromorphic computing *iScience* **23** 101874

Enocean 2015 The true cost of batteries – why energy harvesting is the best power solution for wireless sensors *White Paper* (https://www.enocean.com/fileadmin/redaktion/pdf/white_paper/White_Paper_EnOcean_Cost_of_Batteries.pdf) (Accessed: January 28, 2021)

Fridkin V M 1979 *Photoferroelectrics* 1st edn (Berlin: Springer)

Fu H R, Wang Y G, Guo H, Jain A and Chen F G 2020 Enhancing photostriction in KNN-based ceramics by constructing the morphotropic phase boundary and narrowing the energy band gap *Ceram. Int.* **47** 10996–1002

Glass A M, von der Linde D and Negran T J 1974 High-voltage bulk photovoltaic effect and the photorefractive process in $LiNbO_3$ *Appl. Phys. Lett.* **25** 233

Glass A M, von der Linde D, Auston D H and Negran T J 1975 Excited state polarization, bulk photovoltaic effect and the photorefractive effect in electrically polarized media *J. Electr. Mater.* **4** 915–43

Green M, Dunlop E, Hohl-Ebinger J, Yoshita M, Kopidakis N and Hao X 2021 Solar cell efficiency tables (version 57) *Prog. Photovoltaics Res. Appl.* **29** 3–15

Grinberg I *et al* 2013 Perovskite oxides for visible-light-absorbing ferroelectric and photovoltaic materials *Nature* **503** 509–12

Jella V *et al* 2019 A comprehensive review of flexible piezoelectric generators based on organic-inorganic metal halide perovskites *Nano Energy* **57** 74–93

Kanoun O (ed) 2018 *Energy Harvesting for Wireless Sensor Networks* (Oldenbourg: De Gruyter)

Li F *et al* 2019 Giant piezoelectricity of Sm-doped $Pb(Mg_{1/3}Nb_{2/3})O_3$–$PbTiO_3$ single crystals *Science* **364** 264–8

Li F *et al* 2018 Ultrahigh piezoelectricity in ferroelectric ceramics by design *Nat. Mater.* **17** 349–54

Liu S, Zheng F, Grinberg I and Rappe A M 2016 Photoferroelectric and photopiezoelectric properties of organometal halide perovskites *J. Phys. Chem. Lett.* **7** 1460–5

Lopez-Varo P *et al* 2016 Physical aspects of ferroelectric semiconductors for photovoltaic solar energy conversion *Phys. Rep.* **653** 1–40

Lou Q, Zeng J, Man Z, Zheng L, Park C, Kassiba A, Liu Y, Chen X and Li G 2019 The effect of grain boundary on the visible light absorption of $BaTi_{1-x}[Ni_{1/2}Nb_{1/2}]_xO_{3-\delta}$ ferroelectric ceramics *J. Am. Ceram. Soc.* **102** 7405–13

Moulson A J and Herbert J M 2003 *Electroceramics: Materials, Properties, Applications* 2nd edn (Chichester: Wiley) p 32

Nechache R, Harnagea C, Li S, Cardenas L, Huang W, Chakrabartty J and Rosei F 2015 Bandgap tuning of multiferroic oxide solar cells *Nat. Photonics* **9** 61–7

Nguyen T D, Mao S, Yeh Y W, Purohit P K and McAlpine M C 2013 Nanoscale flexoelectricity *Adv. Mater.* **25** 946–74

Paillard C, Bai X, Infante I C, Guennou M, Geneste G, Alexe M, Kreisel J and Dkhil B 2016 Photovoltaics with ferroelectrics: current status and beyond *Adv. Mater.* **28** 5153–68

Pandey R, Gangadhar S B, Grover S, Singh S K, Kadam A, Ogale S, Waghmare U V, Rao V R and Kabra D 2019 Microscopic origin of piezoelectricity in lead-free halide perovskite: application in nanogenerator design *ACS Energy Lett.* **4** 1004–11

Pérez-Tomas A *et al* 2019 $PbZrTiO_3$ ferroelectric oxide as an electron extraction material for stable halide perovskite solar cells *Sustain. Energy Fuels* **3** 382–9

Rakita Y, Bar-Elli O, Meirzadeh E, Kaslasi H, Peleg Y, Hodes G, Lubomirsky I, Oron D, Ehre D and Cahen D 2017 Tetragonal $CH_3NH_3PbI_3$ is ferroelectric *Proc. Natl Acad. Sci.* **114** E5504–12

Röhm H, Leonhard T, Hoffmann M J and Colsmann A 2020 Ferroelectric poling of methylammonium lead iodide thin films *Adv. Funct. Mater.* **30** 1908657

Rubio-Marcos F, Del Campo A, Ordoñez-Pimentel J, Venet M, Rojas-Hernandez R E, Páez-Margarit D, Ochoa D A, Fernández J F and García J E 2021 Photocontrolled strain in polycrystalline ferroelectrics via domain engineering strategy *ACS Appl. Mater. Interfaces* **13** 20858–64

Rubio-Marcos F, Ochoa D A, Del Campo A, García M A, Castro G R, Fernández J F and García J E 2018 Reversible optical control of macroscopic polarization in ferroelectrics *Nat. Photonics* **12** 29–32

Seidel J, Fu D, Yang S Y, Alarcón-Lladó E, Wu J, Ramesh R and Ager J W III 2011 Efficient photovoltaic current generation at ferroelectric domain walls *Phys. Rev. Lett.* **107** 126805

Shahrokhi S *et al* 2020 Emergence of ferroelectricity in halide perovskites *Small Methods* **4** 2000149

Spanier J E *et al* 2016 Power conversion efficiency exceeding the Shockley–Queisser limit in a ferroelectric insulator *Nat. Photonics* **10** 611–6

Tan Z *et al* 2019 Thinning ferroelectric films for high-efficiency photovoltaics based on the Schottky barrier effect *NPG Asia Mater.* **11** 20

Vats G, Bai Y, Zhang D, Juuti J and Seidel J 2019 Optical control of ferroelectric domains: Nanoscale insight into macroscopic observations *Adv. Opt. Mater.* **7** 1800858

Vats G, Peräntie J, Palosaari J, Juuti J, Seidel J and Bai Y 2020 Current modulation by optoelectric control of ferroelectric domains *ACS Appl. Electron. Mater.* **2** 2829–36

Wu L 2019 Investigation of ferroelectric perovskite oxides for photovoltaic applications *PhD Thesis* University of Pennsylvania

Wu L, Akbashev A R, Podpirka A A, Spanier J E and Davies P K 2019 Infrared-to-ultraviolet light-absorbing $BaTiO_3$-based ferroelectric photovoltaic materials *J. Am. Ceram. Soc.* **102** 4188–99

Würfel P and Würfel U 2016 *Physics of Solar Cells: From Basic Principles to Advanced Concepts* 3rd edn (Weinheim: Wiley)

Xiao H, Dong W, Guo Y, Wang Y, Zhong H, Li Q and Yang M M 2019 Design for highly piezoelectric and visible/near-infrared photoresponsive perovskite oxides *Adv. Mater.* **31** 1805802

Xu X S *et al* 2009 Optical properties and magnetochromism in multiferroic $BiFeO_3$ *Phys. Rev.* B **79** 134425

Yang S Y *et al* 2010 Above-bandgap voltages from ferroelectric photovoltaic devices *Nat. Nanotechnol.* **5** 143–7

Yang Y *et al* 2019 An ultrathin ferroelectric perovskite oxide layer for high-performance hole transport material free carbon based halide perovskite solar cells *Adv. Funct. Mater.* **29** 1806506

Yudin P V and Tagantsev A K 2013 Fundamentals of flexoelectricity in solids *Nanotechnology* **24** 432001

Zhong H, Xiao H, Jiao N and Guo Y 2019 Boosting piezoelectric response of KNN-based ceramics with strong visible-light absorption *J. Am. Ceram. Soc.* **102** 6422–6

Zubko P, Catalan G and Tagantsev A K 2013 Flexoelectric effect in solids *Annu. Rev. Mater. Res.* **43** 387–421

IOP Publishing

Ferroelectrics
Advances in fundamental studies and emerging applications
Yang Bai and Ilya Grinberg

Chapter 6

Topological defects in ferroelectrics

Jan Seidel

Chapters 2 and 5 have discussed the prediction and manipulation of physical properties of ferroelectrics at atomic to lattice level (Ångstrom-scale) from theoretical and experimental aspects, respectively. Chapters 4 and 5 have collaboratively demonstrated functional properties at the macroscopic level (scale of μm–mm), i.e. in ceramics. Topological defects in ferroelectric crystals are found at the intermediate, nanoscale level in ceramics and thin films, with domain walls being the most prominent. The complexity of compounds, lattice asymmetry, phase transitions and boundary conditions lead to new functionality at such nanoscale features. This chapter provides a discussion of topological defects and their properties, including domain walls, vortex/vertex structures and polar skyrmions, including those in emerging van der Waals ferroelectrics. An outlook of this new research aspect of ferroelectrics will be given at the end.

6.1 Topological defects

The arrangement of electrical dipoles in polar materials such as ferroelectrics inevitably leads to the formation of complex polar patterns that involve topological trivial and nontrivial structures. The latter ones are referred to as topological defects (Mermin 1979). Their distinguishing feature is the fact that they carry a topological charge (Chern number), and that they cannot be easily removed from an ordered system, unless it is finite and boundaries break the symmetry, as for example in thin film ferroelectrics. One of the most studied topological defects in ferroelectrics are domain walls (Seidel 2016). These exist in various forms and configurations with a wide variety of distinguishing properties that have been envisioned for example in nanoelectronic applications (Catalan *et al* 2012). More recently, nanoscale vortex (or vertex) patterns and polar skyrmions have also been fabricated and observed in ferroelectrics. These types of structures typically require elaborate engineering of the associated materials in island, multilayer and heterostructure geometries. Many of the classic ferroelectric systems, although not all, are perovskites and oxide

doi:10.1088/978-0-7503-3975-9ch6

materials. More recently, van der Waals layered ferroelectric systems that can even be engineered out of non-ferroelectric starting materials have emerged. These latest developments are likely to bring about new opportunities for 'domain wall engineering' and the exploitation of functional topological defects in these material systems. While some of these research areas are just starting out and considerable work is to be done, selected aspects regarding topological defects are discussed in this chapter.

6.2 Domain walls

6.2.1 General properties

Domains in ferroelectrics form as a consequence of energy minimization associated with electrostatic boundary conditions. Further reasons for their formation include nonuniform strain because strain and polar order are usually coupled in such systems and they strongly influence each other, as well as microscopic and nanoscopic defects, such as grain boundaries, phase boundaries, vacancies, interstitials, etc. Additional influences arise from the thermal and electrical poling history of the sample.

In all ferroelectrics the macroscopic bulk polarization is formed by a displacement of charge, positive relative to negative, at the unit cell level of the material. Inhomogeneities in the polar order, therefore, can be treated as a local charge density variation, which exist at surfaces, boundaries or interfaces, for example. Material systems will minimize their energy by eliminating these charges as far as possible. In thin films, for example, domains are formed that involve flux-closure geometries to minimize energetically costly stray fields. This, however, comes at the cost of creating additional domain walls, which are not the ground state of the polarized system, so overall a balance is found in equilibrium. Additionally, because polarization is often coupled to strain, local strain fields are involved in the overall energetic picture. In recent years, ferroelectric domain walls have been characterized to exhibit non-Ising-like polarization reorientation, consisting of mixtures of Néel-type, Bloch-type, and Ising type, making their intrinsic properties rich and interesting (Rodriguez *et al* 2009, Wei *et al* 2016, Cherifi-Hertel 2017).

The above considerations make it obvious that only certain configurations of domain structures and associated domain walls are allowed unless they are brought about or interact with additional defects or boundary conditions. These defects, i.e. impurities and migrating ions, can also help to screen surface and interface polarization in ferroelectrics. Considering activation energies and migration at room temperature, these processes can be fairly slow in oxide perovskites. The influence on domain wall properties has recently been considered and mechanisms for memory and device applications have been discussed.

6.2.2 Functionality in nanoelectronics

Over the past decade, domain walls in ferroelectrics have seen considerable research attention because of their interesting electronic properties and associated functionality. The potential for creating reconfigurable active nanoelectronics elements with intrinsic material properties at the domain wall, that are different from the bulk material itself, has been discussed in terms of changes in structure and following electronic structure, which can be probed with high-resolution microscopy and spectroscopy methods. One of the main driving factors has been the observation of domain wall conductivity in semiconducting and insulating bulk systems. Initial observations of the phenomenon in materials such as $BiFeO_3$ (BFO) and $YMnO_3$ were quickly followed by a large array of other materials and in fact many of the otherwise well-known classical ferroelectric systems and newly synthesized materials.

Control of these electronic domain wall properties by additional doping and strain engineering also deserves a closer look. A prototypical example can be found for the material $BiFeO_3$, which has been widely studied. Generic doping approaches to alter the material functionality include A- and B-site substitutions of the perovskite lattice to alter electronic properties, but also magnetic properties in this multiferroic system. Doping with Ca for example generates large amounts of oxygen vacancies to achieve charge neutrality in the material. These in turn interact with domain walls and alter optical properties of the material. Other examples include doping with La, which initially was investigated in conjunction with varying oxygen partial pressure during growth of the material and its influence on domain wall conductivity. In many cases this conductivity was shown to be thermally activated with activation energies in the range of 0.2–0.5 eV, although examples of metallic conduction behaviour in other ferroelectrics have been reported as well. More recently, certain types of hybrid domain walls or structural phase boundaries in morphotropic systems have also been investigated and showed similar behaviour to conventional domain walls in BFO.

Recent demonstrations of functional elements based on domain walls include memory cells and memristive elements, among others. In all cases the basic function of the element is produced by the domain wall, its controllable presence or absence, or controlled alteration of intrinsic properties by electric fields, as shown in figure 6.1. Other applications include the targeted alteration of intrinsic polarization in ferroelectrics and their stability and retention through domain wall engineering. One recent example is the demonstrated designer-defect-engineered mediated polarization retention with virtually no degradation in switched nanoscale domains of epitaxial BFO films for periods of several years, as demonstrated in figure 6.2 (Zhang *et al* 2020, 2021a). This represents a significant improvement over the best values hitherto reported. Scanning probe microscopy-based dynamic switching measurements reveal a significantly increased activation field for domain wall movement in such materials. Atomic resolution scanning transmission electron microscopy indicates that nanoscale defect pockets pervade the entire film thickness. These defects act as highly efficient domain wall pinning centres, resulting in anomalous retention. Findings of the author's team demonstrate that defects can be exploited in a positive manner to solve reliability issues in ferroelectric films used in functional devices.

Domains and domain walls can also be manipulated by light illumination. Light-induced changes in ferroelectrics have been discussed in conjunction with (i) localized heating assisted diffusion of alkali element (as in $LiNbO_3$) and oxygen

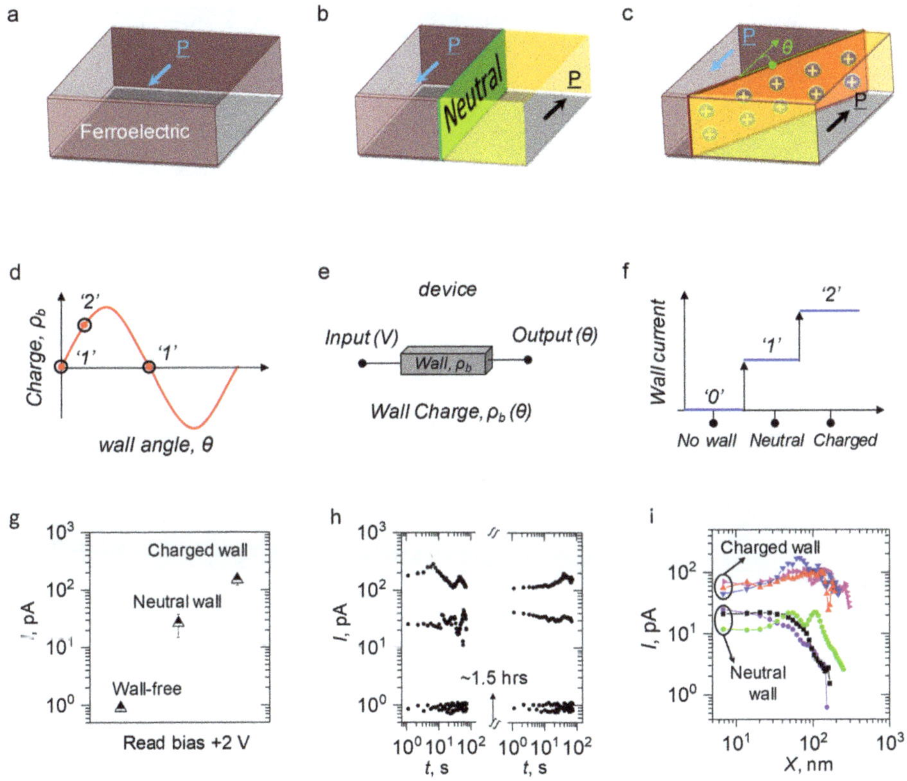

Figure 6.1. Concept of a multistate conformational domain wall switch. (a)–(c) Ferroelectric material in the absence of domain wall: monodomain state, or off-state (a), in the presence of a neutral (b), and charged (c) domain walls, respectively. (d) A plot showing evolution of polarization charge on domain wall plane as a function of its in-plane angle (θ). (e) A domain wall device with a variable output controlled by conformation of wall (θ). (f) Three distinct states of such a device corresponding to the following wall conformations: no wall, neutral wall, and charged wall. (g) A plot of the measured electrical current between the electrodes for three domain wall configurations. (h) Electrical current versus time for three domain wall states. There is a time lapse (break) of ~1.5 h between left-half and right-half of the plot. (i) Profile of electrical conduction for the neutral and charged domain walls. Reproduced with permission from Sharma *et al* (2019b). Copyright © 2019 John Wiley & Sons.

vacancies (Steigerwald 2010), (ii) changes in oxidation state of the central atom (e.g. in $BaTiO_3$) (Abel *et al* 2013) and (iii) Jahn–Teller distortions due to light-induced pyroelectric current or charge injection in the conduction band (as in SbSI) (Fridkin *et al* 1967, Fridkin 1979). Warren *et al* (1995) suggested that similar to electrical and thermal fluctuations, an exposure to light involves 'locking domains by electronic charge trapping at domain boundaries'. Recently, light dependent nanoscale changes in KNN–BNNO (Ba A-site and Ni B-site doped $(K,Na)NbO_3$), as illustrated in figure 6.3 (Vats *et al* 2019, 2020a, 2020b, 2021, Bai *et al* 2018), have been investigated using photoelectric atomic force microscopy. In this study, piezoresponse force microscopy (PFM) and Kelvin probe force microscopy (KPFM) measurements have been performed and the obtained results have been

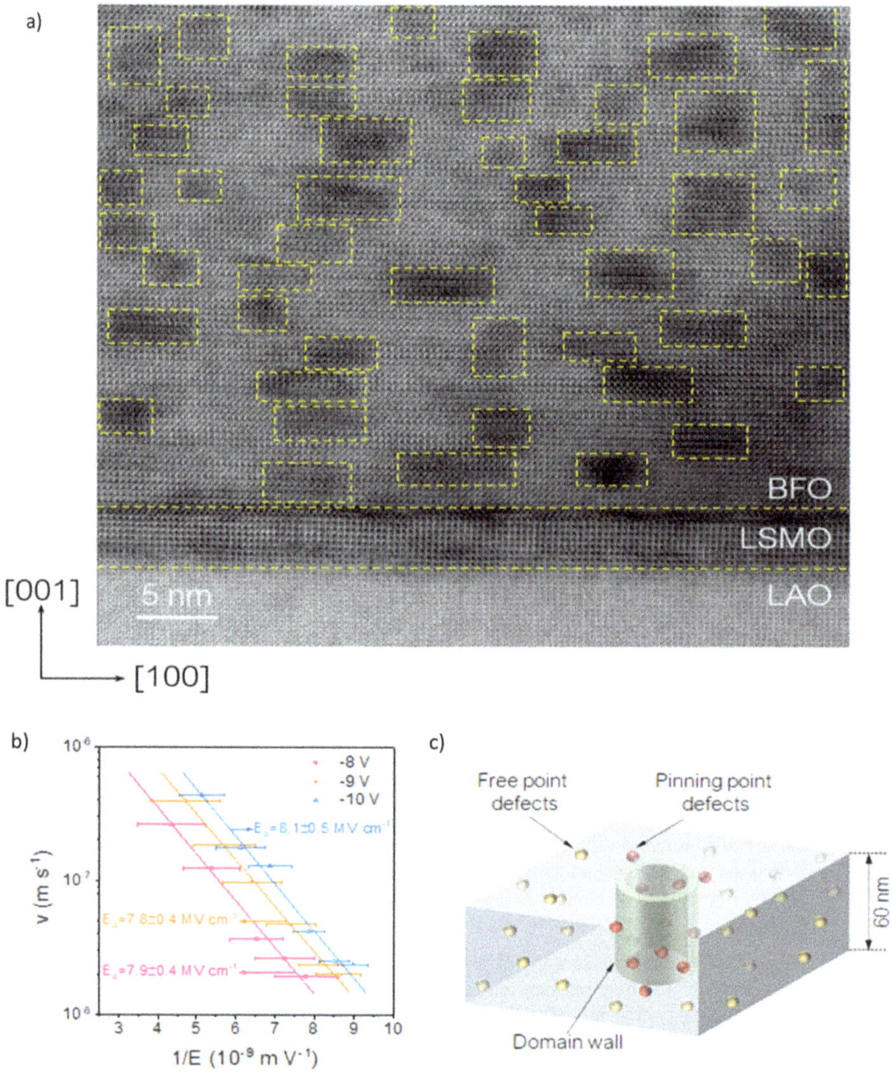

Figure 6.2. Designer defect (a) engineering of ferroelectric properties and resulting domain wall pinning showing world-record domain wall activation energies (b). (c) Sketch of the pinning mechanism. Reproduced under the CC-BY license from Zhang *et al* (2020).

correlated with macroscopic observations. This work points at the possibility of developing novel device mechanisms to achieve optical control over ferroelectricity and ferroelectric domain walls by effective utilization of the associated semiconducting material characteristics.

The observed and demonstrated functionality of domain walls as described above has opened new pathways for their application in nanoelectronic circuits and devices. Pertaining challenges for a broader uptake have been discussed in detail recently.

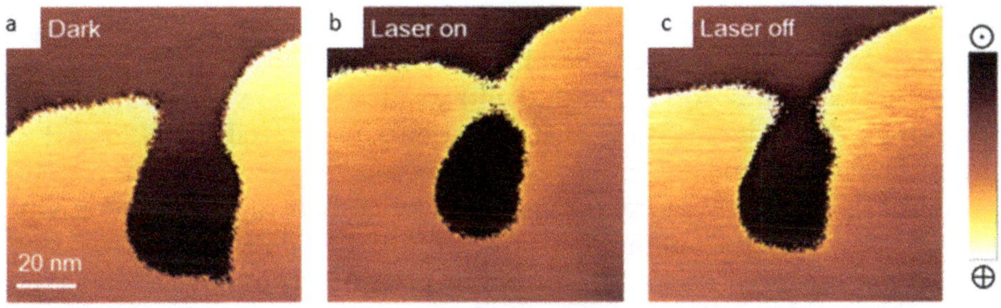

Figure 6.3. Light-induced domain wall manipulation in ferroelectric KNN–BNNO. (a)–(c) The magnified view of domain wall motion corresponding to dark, laser exposure and post laser exposure state. Reproduced with permission from Vats *et al* (2019). Copyright © 2019 John Wiley & Sons.

6.3 Vortex and vertex structures

6.3.1 Patterns in nanoislands and heterostructures

Naturally forming vortex-like vertex patterns exist in certain types of ferroelectrics, for example in the improper hexagonal manganite systems (Meier *et al* 2012). Engineered patterns of domain structures with targeted domain wall meeting points were also investigated in conventional ferroelectrics, for example, by Balke *et al* (2012). In this work electronic conductivity at the vortex-like meeting point of domain walls was shown. More recent work translates this concept to engineered oxide heterostructures (Yadav *et al* 2016) and nanoscale island structures, which provide another level of control over the domain formation. More complex polar topological defects and structures can be built on this basis. These include flux-closure domain structures (Ivry *et al* 2010, McQuaid *et al* 2011), centre-type domains (Li *et al* 2017), labyrinth-like domain patterns (Nahas *et al* 2020a), and curl and spiral patterns (Peters *et al* 2016, Choudhury *et al* 2011).

Several groups have reported the controlled generation and manipulation of vortex and vertex structures in such systems. Recently, exotic crossed quadrant centre convergent and divergent domain structures that enable the creation of head-to-head and tail-to-tail domain wall configurations (Ma *et al* 2018, Kim *et al* 2018a, 2018b), respectively, were reported in nanostructures of BFO. These structures were produced by templating or self-assembled growth and ranged in size from tens of nanometers to hundreds of nanometers. Local configurations of domains and domain wall types were studied using vector PFM and high-resolution transmission electron microscopy. The enhancement of the domain wall currents in such structures is shown to be due to the formation of conducting pathways of charged defects accumulated along domain walls and traversing the thickness of the nano-crystals. The diverse domain walls could be manipulated by an applied electric field in various architectures.

6.3.2 Manipulation and properties

The engineered domain patterns in such nanoislands are radially convergent and divergent states that are electrically addressable. Electrical conductivity at walls was shown to be tunable by several orders of magnitude in several cases (Yang *et al* 2021). The observation of highly conductive domain walls in so-called quadrant centre topological states, formed in the self-assembled square BFO nanoislands, as shown in figure 6.4, is an example. Interestingly, the specific topological states (centre divergent and convergent states) can be controllably switched, resulting in high control of the centre current, which can be utilised for information storage. Possible memory devices with nondestructive current readout have been proposed.

Strain engineering of ferroelectric oxide heterostructures has led to widely investigated vortex lattices consisting of clockwise and anticlockwise vortices, as first shown in $SrTiO_3/PbTiO_3$ (STO/PTO). These patterns are associated with a recently discussed unique negative capacitance in the PTO layers (Yadav *et al* 2019). The current explanation is that the high energy cores of the vortices suppress the local electric polarization, resulting in the overall macroscopic negative capacitance. This property has been discussed in terms of lowering the operation voltages of field effect transistor structures and eventually reducing the energy consumption of such devices.

Control over the vortex patterns has been demonstrated to some degree, for example, polar vortex pair splitting and a phase transformation from the vortex to a polar wave structure, which can be triggered by an external bias (Du *et al* 2019). Optical and thermal control was achieved by light illumination and heating, which induced a topological transition from a vortex-a1/a2 mixed phase to a 3D periodical topological domain array in STO/PTO superlattices (Stoica *et al* 2019). In addition, mechanical control of patterns and domain structures was also demonstrated, driven by applied mechanical force (Chen *et al* 2020).

Vortex arrays in the STO/PTO superlattice have also been investigated with respect to their optical properties, for example, in optical second-harmonic

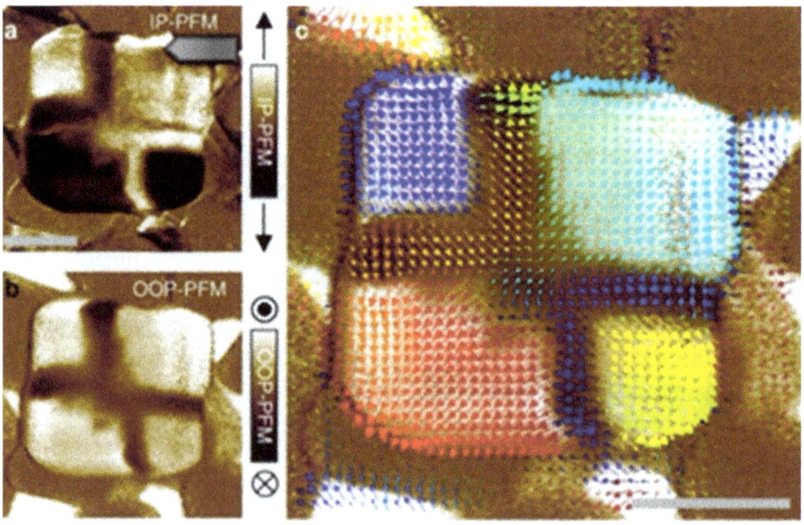

Figure 6.4. Vertex textures in strain graded ferroelectric BFO nanoplates. Reproduced under the CC-BY license from Kim *et al* (2018a).

generation (SHG) experiment. Work by Damodaran *et al* (2017) investigated gyrotropic electrotoroidal of such structures and suggested further control of such properties through control of the topological patterns. Another interesting aspect of these engineered vortex structures is the emergence of toroidal order. Applying an electric field enables investigations of coupling between the toroidal and ferroelectric order parameters in such materials. This concept can be used to control other properties, for example chirality. Recent resonant soft x-ray diffraction studies have shown that the polar vortices in PTO/STO superlattices exhibit strong circular dichroism, hinting at chiral structures (Warren *et al* 1995). Additional circular dichroism (XCD) measurements (Shafer *et al* 2018) are further strongly indicative of chirality.

The discussed vortex-like topological states enable the engineering of materials for developing high performance devices, for example in memory and information processing.

6.4 Polar skyrmions

6.4.1 Strain engineering in heterostructures

Skyrmions in solid-state materials were first observed in magnetic systems, and more recently also in ferroelectrics (Li *et al* 2018, Nahas *et al* 2015, Das *et al* 2019). The presence of polar skyrmions in ferroelectrics was investigated widely by theoretical considerations, for example in bubbles in PTO/STO superlattice structures. The influence of external electric field application using phase-field simulations (Hong and Chen 2018) and second-principles simulations (Goncalves *et al* 2019), was discussed as an approach to create polar skyrmions by strain engineering of the ferroelectric $PbTiO_3$ layer in heterostructures. Initial discussions of the control and manipulation of such polar skyrmions by various stimuli, including temperature, electric field and epitaxial strain, were also discussed.

Das *et al* (2019) then reported on the experimental observation of skyrmion bubbles in PTO/STO heterostructures by varying epitaxial constraints that has been the platform for a wide variety of studies of such topological defects by several groups.

6.4.2 Emerging functionality

Skyrmions in solids can be considered as ultra-small elemental building blocks for future integrated information devices. The development of functionality is several years ahead of that in ferroelectrics and follows initial observations of such whirl-like spin textures that has been observed in chiral magnets (Muhlbauer *et al* 2009, Yu *et al* 2010) and multiferroic materials, for example Cu_2OSeO_3 (Seki *et al* 2012, Han *et al* 2020). Information processing in memory and logic devices has been investigated in such materials, including high frequency (GHz–THz range) manipulation for nanoelectronic devices. Similar applications can be envisioned for ferroelectrics. Local alteration of properties can also be found, for example, in a giant enhancement of piezoelectric response in individual polar skyrmion-like textures (Zhuo and Yang 2020). The creation and annihilation of skyrmion-like bubble states were also both demonstrated in PZT ($Pb(Zr,Ti)O_3$) ultrathin films and heterostructures (Zhang *et al* 2017, Nahas *et al* 2020b).

6.5 Topological defects in van der Waals ferroelectrics

A relatively recent addition to the arsenal of ferroelectric materials is layered van der Waals (vdW) ferroelectrics. These include materials such as In_2Se_3 (Zheng *et al* 2018), CIPS (Zhang *et al* 2021b), and WTe_2 (Sharma *et al* 2019a). The layered crystal structure of van der Waals ferroelectrics combining strong intra- and weak interlayer forces leads to unique properties that are not found in more conventional ferroelectric materials. Fundamentally new phenomena in vdW ferroelectric materials include downscaling limits, polarization origins, and novel switching mechanisms, as well as unique features such as quadruple-well potentials, metallic ferroelectricity, and dipole locking effects (Zhang *et al* 2022). Additionally, because of their easily stackable nature, vdW ferroelectrics are relatively easy to integrate with highly dissimilar materials (such as CMOS (complementary metal-oxide-semiconductor) silicon) without significant interfacial issues otherwise found. This makes them attractive building blocks for future nanoelectronics.

A recent development highlight is the emerging field of engineered vdW ferroelectricity, which is created by artificially broken centrosymmetry in stacks of otherwise nonpolar parent materials (Yasuda *et al* 2021, Tsymbal 2021). This concept opens a wide variety of possible new designer ferroelectrics that can be utilized for Boltzmann-tyranny beating transistors, nonvolatile memory, optoelectronic and flexible devices, to name only a few possible applications. Recent progress is fast-paced and points to new opportunities in technology and future application development.

WTe_2 has recently been shown to be ferroelectric despite being a semimetal. Usually these properties are mutually exclusive, although the coexistence of reversible polar distortions and metallicity leading to a ferroelectric metal was suggested by Anderson and Blount (1965). Clear domain structures and switching of the polarization can be observed in figure 6.5. The observed ferroelectric domains in thick bulk crystals form in different shapes and ferroelectric switching in ~15 nm films shows typical ferroelectric hysteretic behaviour. The question of the physical extent of these domains is an interesting one. One possibility is that the piezoresponse and ferroelectric domains come only from within a Debye length of the surface. However, it is also possible that the ferroelectric domains extend much deeper into the material, since ferroelectricity and ferroelasticity are often coupled in conventional ferroelectrics and as ferroelastic domains have been predicted for bulk WTe_2 (Li and Li 2016). Little is known so far about the domain walls in such metallic ferroelectric systems. The question of screening around these topological defects by a metallic background could lead to new functionality.

6.6 Summary and outlook

The potentially perceived negative connotation of 'defect' in topological defects is a misnomer given their widely investigated special properties over the last years.

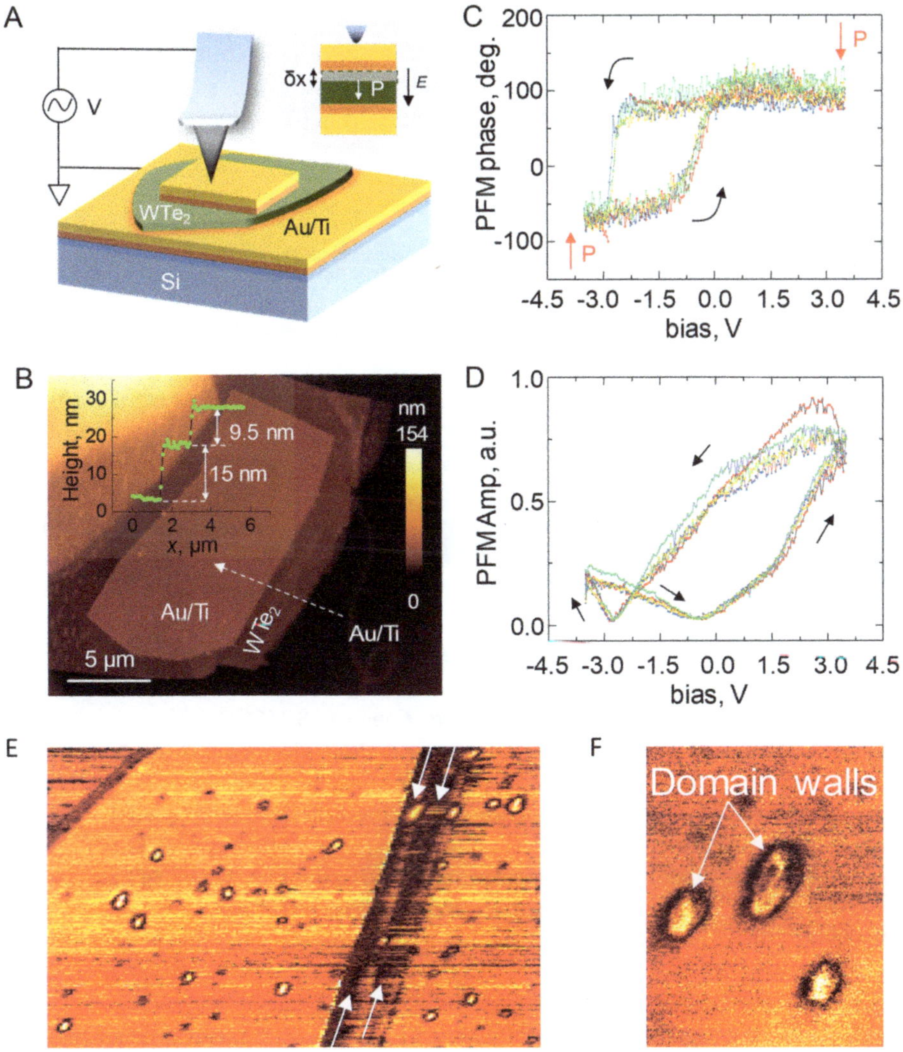

Figure 6.5. Probing ferroelectricity in WTe_2 thin film samples (Sharma *et al* 2019a). (a) Schematic of the experimental set-up and geometry. Inset in (a) shows zoomed-in side view of the device structure. (b) Topography image of the metal (Ti/Au)-encapsulated WTe_2 flake on Si/SiO_2 substrate covered with a conductive buffer layer of Ti/Au. Inset in (b) shows the profile of surface morphology along dashed white-line. (c and d) Spectroscopic bias-dependent piezoresponse phase (c), and amplitude (d) hysteretic curves acquired through the top metal electrode gating the WTe_2 flake. (e and f) Ferroelectric domains seen in PFM amplitude. Reproduced with permission (Sharma *et al* 2019a). Copyright © 2019 Science.

A more appropriate term would be 'topological assets'. Their small sizes in the nanometer range make them challenging to study, nevertheless, some significant inroads have been made recently and initial demonstrations of prototype devices and functionality have been published. New material developments are again likely to bring about new research directions, as is the case with van der Waals ferroelectrics, for example. The study of topological defects has not only become a significant subfield of the investigation of ferroelectric materials, but also has received considerable attention in other related material classes, for example multiferroics and magnetic materials. Overall, this research field is likely to see increased activity in the coming years.

References

Abel S *et al* 2013 A strong electro-optically active lead-free ferroelectric integrated on silicon *Nat. Commun.* **4** 1671

Anderson P W and Blount E 1965 Symmetry considerations on martensitic transformations: "ferroelectric" metals? *Phys. Rev. Lett.* **14** 217

Bai Y, Vats G, Seidel J, Jantunen H and Juuti J 2018 Boosting photovoltaic output of ferroelectric ceramics by optoelectric control of domains *Adv. Mater.* **30** 1803821

Balke N *et al* 2012 Enhanced electric conductivity at ferroelectric vortex cores in $BiFeO_3$ *Nat. Phys.* **8** 81–8

Catalan G, Seidel J, Ramesh R and Scott J F 2012 Domain wall nanoelectronics *Rev. Mod. Phys.* **84** 119

Chen P *et al* 2020 Atomic imaging of mechanically induced topological transition of ferroelectric vortices *Nat. Commun.* **11** 1840

Cherifi-Hertel S, Bulou H, Hertel R, Taupier G, Dorkenoo K D H, Andreas C, Guyonnet J, Gaponenko I, Gallo K and Paruch P 2017 Non-ising and chiral ferroelectric domain walls revealed by nonlinear optical microscopy *Nat. Commun.* **8** 15768

Choudhury N, Walizer L, Lisenkov S and Bellaiche L 2011 Geometric frustration in compositionally modulated ferroelectrics *Nature* **470** 513

Damodaran A R *et al* 2017 Phase coexistence and electric-field control of toroidal order in oxide superlattices *Nat. Mater.* **16** 1003

Das S *et al* 2019 Observation of room temperature polar skyrmions *Nature* **568** 368–72

Du K, Zhang M, Dai C, Zhou Z N, Xie Y W, Ren Z H, Tian H, Chen L Q, Tendeloo G V and Zhang Z 2019 Emerging phenomena from exotic ferroelectric topological states *Nat. Commun.* **10** 4864

Fridkin V M 1979 *Photoferroelectrics* (Berlin: Springer)

Fridkin V, Groshik I, Lakhovizkaya V, Mikhailov M and Nosov V 1967 Current saturation and photoferroelectric effect in SbSI *Appl. Phys. Lett.* **10** 354–6

Goncalves M A P, Escorihuela-Sayalero C, Garca-Fernandez P, Junquera J and Iniguez J 2019 Theoretical guidelines to create and tune electric skyrmion bubbles *Sci. Adv.* **5** eaau7023

Han M G *et al* 2020 Scaling, rotation and channelling behaviour of helical and skyrmion spin textures in thin films of Te-doped Cu_2OSeO_3 *Sci. Adv.* **6** eaax2138

Hong Z and Chen L Q 2018 Blowing polar skyrmion bubbles in oxide superlattices *Acta Mater.* **152** 155

Ivry Y, Chu D P, Scott J F and Durkan C 2010 Flux closure vortexlike domain structures in ferroelectric thin films *Phys. Rev. Lett.* **104** 207602

Kim K E *et al* 2018a Configurable topological textures in strain graded ferroelectric nanoplates *Nat. Commun.* **9** 403

Kim K E, Kim Y J, Zhang Y, Xue F, Kim G Y, Song K, Choi S Y, Liu J M, Chen L Q and Yang C H 2018b Ferroelastically protected polarization switching pathways to control electrical conductivity in strain-graded ferroelectric nanoplates *Phys. Rev. Mater.* **2** 084412

Li L, Cheng X, Jokisaari J R, Gao P, Britson J, Adamo C, Heikes C, Schlom D G, Chen L Q and Pan X 2018 Defect-induced hedgehog polarization states in multiferroics *Phys. Rev. Lett.* **120** 137602

Li W and Li J 2016 Ferroelasticity and domain physics in two-dimensional transition metal dichalcogenide monolayers *Nat. Commun.* **7** 10843

Li Z *et al* 2017 High-density array of ferroelectric nanodots with robust and reversibly switchable topological domain states *Sci. Adv.* **3** e1700919

Ma J *et al* 2018 Controllable conductive readout in self-assembled, topologically confined ferroelectric domain walls *Nat. Nanotechnol.* **13** 947–52

McQuaid R G P, McGilly L J, Sharma P, Gruverman A and Gregg J M 2011 Mesoscale flux-closure domain formation in single-crystal $BaTiO_3$ *Nat. Commun.* **2** 404

Meier D, Seidel J, Cano A, Delaney K, Kumagai Y, Mostovoy M, Spaldin N A, Ramesh R and Fiebig M 2012 Anisotropic conductance at improper ferroelectric domain walls *Nat. Mater.* **11** 284

Mermin N D 1979 The topological theory of defects in ordered media *Rev. Mod. Phys.* **51** 591–8

Muhlbauer S, Binz B, Jonietz F, Pfleiderer C, Rosch A, Neubauer A, Georgii R and Boni P 2009 Skyrmion lattice in a chiral magnet *Science* **323** 915

Nahas Y, Prokhorenko S, Louis L, Gui Z, Kornev I and Bellaiche L 2015 Discovery of stable skyrmionic state in ferroelectric nanocomposites *Nat. Commun.* **6** 8542

Nahas Y *et al* 2020a Inverse transition of labyrinthine domain patterns in ferroelectric thin films *Nature* **577** 47

Nahas Y, Prokhorenko S, Zhang Q, Govinden V, Valanoor N and Bellaiche L 2020b Topology and control of self-assembled domain patterns in low-dimensional ferroelectrics *Nat. Commun.* **11** 5779

Peters J J P, Apachitei G, Beanland R, Alexe M and Sanchez A M 2016 Polarization curling and flux closures in multiferroic tunnel junctions *Nat. Commun.* **7** 13484

Rodriguez B J, Gao X S, Liu L F, Lee W, Naumov I I, Bratkovsky A M, Hesse D and Alexe M 2009 Vortex polarization states in nanoscale ferroelectric arrays *Nano Lett.* **9** 1127

Seidel J (ed) 2016 *Topological Structures in Ferroic Materials* (Berlin: Springer)

Seki S, Ishiwata S and Tokura Y 2012 Magnetoelectric nature of skyrmions in a chiral magnetic insulator Cu_2OSeO_3 *Phys. Rev.* B **86** 060403

Shafer P *et al* 2018 Emergent chirality in the electric polarization texture of titanate superlattices *Proc. Natl Acad. Sci. USA* **115** 915

Sharma P, Xiang F X, Shao D F, Zhang D, Tsymbal E Y, Hamilton A R and Seidel J 2019a A room-temperature ferroelectric semimetal *Sci. Adv.* **5** eaax5080

Sharma P *et al* 2019b Conformational domain wall switch *Adv. Funct. Mater.* **29** 1807523

Steigerwald H, Lilienblum M, Von Cube F, Ying Y, Eason R, Mailis S, Sturman B, Soergel E and Buse K 2010 Origin of UV-induced poling inhibition in lithium niobate crystals *Phys. Rev.* B **82** 214105

Stoica V A *et al* 2019 Optical creation of a supercrystal with three-dimensional nanoscale periodicity *Nat. Mater.* **18** 377

Tsymbal E Y 2021 Two-dimensional ferroelectricity by design *Science* **372** 1389–90

Vats G, Bai Y, Zhang D, Juuti J and Seidel J 2019 Optical control of ferroelectric domains: Nanoscale insight on macroscopic observations *Adv. Opt. Mater.* **7** 1800858

Vats G, Peräntie J, Palosaari J, Juuti J, Seidel J and Bai Y 2020a Current modulation by opto-electric control of ferroelectric domains *ACS Appl. Electron. Mater* **2** 2829

Vats G, Peräntie J, Juuti J, Seidel J and Bai Y 2020b Coalition of thermo-opto-electric effects in ferroelectrics for enhanced cyclic energy conversion *Energy Technol* **8** 2000500

Vats G, Bai Y and Seidel J 2021 Opto-mechanical mapping of ferroelectric domains and the piezo-photovoltaic effect in Ba- and Ni-doped $(K_{0.5}Na_{0.5})NbO_3$ *Adv. Photonics Res* **2** 2100050

Warren W L, Dimos D, Tuttle B, Pike G, Schwartz R, Clews P and McIntyre D 1995 Polarization suppression in $Pb(Zr,Ti)O_3$ thin films *J. Appl. Phys.* **77** 6695–702

Wei X K, Jia C L, Sluka T, Wang B X, Ye Z G and Setter N 2016 Néel-like domain walls in ferroelectric $Pb(Zr,Ti)O_3$ single crystals *Nat. Commun.* **7** 12385

Yadav A K *et al* 2016 Observation of polar vortices in oxide superlattices *Nature* **530** 198–201

Yadav A K *et al* 2019 Spatially resolved steady-state negative capacitance *Nature* **565** 468–71

Yang W D *et al* 2021 Quasi-one-dimensional metallic conduction channels in exotic ferroelectric topological defects *Nat. Commun.* **12** 1306

Yasuda K, Wang X, Watanabe K, Taniguchi T and Jarillo-Herrero P 2021 Stacking-engineered ferroelectricity in bilayer boron nitride *Science* **372** 1458–62

Yu X Z, Onose Y, Kanazawa N, Park J H, Han J H, Matsui Y, Nagaosa N and Tokura Y 2010 Real-space observation of a two-dimensional skyrmion crystal *Nature* **465** 901

Zhang D, Sando D, Sharma P, Cheng X, Ji F, Govinden V, Weyland M, Nagarajan V and Seidel J 2020 Superior polarization retention through engineered domain wall pinning *Nat. Commun.* **11** 349

Zhang D, Sando D, Pan Y, Sharma P and Seidel J 2021a Robust ferroelectric polarization retention in harsh environments through engineered domain wall pinning *J. Appl. Phys.* **129** 014102

Zhang D, Luo Z D, Yin Y, Schoenherr P, Sha C, Pan Y, Sharma P, Alexe M and Seidel J 2021b Anisotropic ion migration and electronic conduction in van der Waals ferroelectric $CuInP_2S_6$ *Nano Lett.* **21** 995

Zhang D, Schoenherr P, Sharma P and Seidel J 2022 Ferroelectric order in van der Waals layered materials *Nat. Rev. Mater* (https://doi.org/10.1038/s41578-022-00484-3)

Zhang Q, Xie L, Liu G, Prokhorenko S, Nahas Y, Pan X, Bellaiche L, Gruverman A and Valanoor N 2017 Nanoscale bubble domains and topological transitions in ultrathin ferroelectric films *Adv. Mater.* **29** e1702375

Zheng C *et al* 2018 Room-temperature in-plane ferroelectricity in van der Waals In_2Se_3 *Sci. Adv.* **4** eaar7720

Zhuo F P and Yang C H 2020 Observation of a stable fractionalized polar skyrmionlike texture with giant piezoelectric response enhancement *Phys. Rev.* B **102** 214112

IOP Publishing

Ferroelectrics

Advances in fundamental studies and emerging applications

Yang Bai and Ilya Grinberg

Chapter 7

Ferroelectricity in halide perovskites

Holger Röhm

Classic ferroelectrics are mostly oxide perovskites, some of which have been discussed in the previous chapters. In contrast, halide perovskites usually possess semiconducting and ionic conducting properties, and thus, examining the potential ferroelectricity is complicated. The susceptibility to degradation also impedes or limits many established characterization techniques known from ceramic ferroelectrics. These factors have contributed to the controversy in the scientific community regarding ferroelectricity in halide perovskites. This chapter provides an overview on the material class of halide perovskites, their novel applications in thin film solar cells, and how ferroelectricity affects their properties. Novel measurement approaches reveal material properties related to ferroelectricity, providing not only insights into the working principles of halide perovskite devices but also hint towards pitfalls for common characterization techniques.

7.1 The rise of halide perovskites and their applications

7.1.1 Halide perovskites

Halide perovskites can refer to a multitude of crystalline compounds, yet in recent years this term is often used synonymously for a class of organic or inorganic metal halides with semiconducting properties. These materials have found a variety of applications in optoelectronics, particularly in photovoltaics (Green *et al* 2014). In 2006, Kojima *et al* for the first time reported the use of methylammonium lead bromide as a light absorber in dye sensitized solar cells (Kojima *et al* 2006). While these initial devices were only short-lived, their promising power conversion efficiency (PCE) of 2.2% soon lead to the discovery of solid state solar cells that no longer required any liquid electrolyte, hence marking the birth of what is nowadays known as perovskite solar cells (Lee *et al* 2012, Kim *et al* 2012).

Within halide perovskite semiconductors, we can differentiate between three general types based on the A-site cation of the ABX_3 structure while all three types mainly utilise Pb or Sn at the B site, and Br, Cl or I on the X site:

doi:10.1088/978-0-7503-3975-9ch7

(1) Organic metal halide (OMH) perovskites, which contain a small cation molecule at the A-site;
(2) Inorganic metal halides (IMH) most often comprising Cs at the A-site of the crystal and;
(3) Solid solutions of types (1) and (2).

The most common A-site cations used in OMH perovskites are methylammonium (MA) and formamidinium (FA) while small amounts of other similarly sized cation molecules can be tolerated by the crystal lattice (Hsieh *et al* 2017, Wang *et al* 2017a). There is an ongoing discussion whether these dipolar organic molecules are in fact required for the superb electronic properties that allowed these materials to facilitate thin film solar cells with PCEs beyond 25%. Some simulations suggested that alignment of these dipoles could influence photogenerated free charge carriers while others rather concluded minimal or no influence (Frost and Walsh 2016, Pecchia *et al* 2015, Wilson *et al* 2019). One of the main reasons for the multitude of optoelectronic applications of halide perovskites are their long-lived electronic charge carriers based on low nonradiative recombination rates that minimize thermal losses. These properties would not be unusual in perfect single crystalline materials as is known from Si or GaAs, but halide perovskites provide these properties in polycrystalline thin films that are commonly deposited from precursor solutions and therefore comprise high defect densities (Wang *et al* 2018, Ono *et al* 2019).

Generally, both organic molecules and halides limit the temperature stability of halide perovskites hence driving the search for new compositions. So far, both inorganic and organic halide perovskites could not reach the same PCEs in solar cells unless a significant portion of the A-site comprised either MA or FA molecules, the B site comprised mostly Pb^{2+} ions and the X site comprised Br, Cl or I (Tai *et al* 2019, Chen *et al* 2021). Figure 7.1 contains an overview on the band gaps as well as valence and conduction band energies of the most common metal halide compositions.

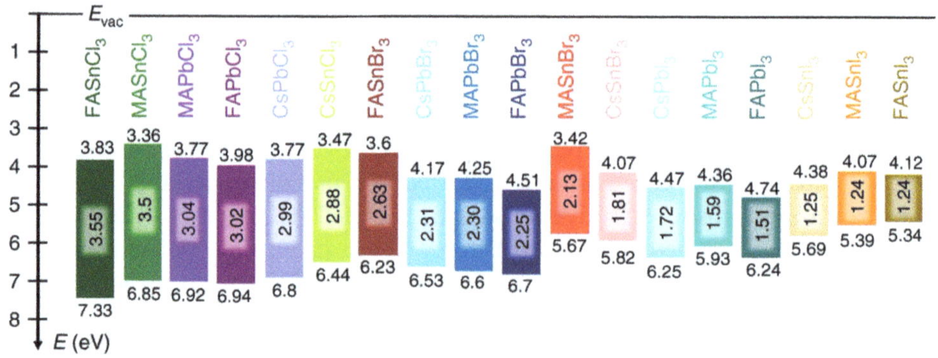

Figure 7.1. Compositional range for band gap tuning of halide perovskites using Cs^+, FA^+ and MA^+ as an A-site cation, Pb^{2+} or Sn^{2+} as a B-site cation and Br^-, Cl^- or I^- as an X-site anion. Reproduced according to CC BY-SA 4.0, Tao *et al* (2019).

In order to tune the optical and electronic properties of halide perovskites and to increase their long-term stability, the use of solid solutions became a common approach. This approach allows design of halide perovskite materials with a band gap ranging from 1.1 eV up to 4.5 eV. This is pivotal for tandem solar cells, where a wide band gap solar cell is combined with a small band gap solar cell in order to overcome the Shockley–Queisser limit for single junction solar cells and thereby enables PCEs of more than 33%. Furthermore, changing the band gap enables colour tuning for light emitting devices. Currently, most high PCE perovskite solar cells use solid solutions of $(Cs,FA,MA,)Pb(I,Br)_3$ with small concentrations (<20 mol%) of Cs and Br (Saliba *et al* 2016, Yoo *et al* 2021). Besides the aforementioned regular 3D perovskite structures, the use of 2D or 1D perovskites turned out to provide a wide range of material properties, particularly used as interface modification for 3D perovskite absorbers (Chao *et al* 2018).

7.1.2 Common applications for halide perovskites

The first major application for halide perovskite semiconductors was their use as light absorbing 'active' material in solar cells that led to their rise within the global research community. Based on their optoelectronic properties such as the sharp absorption/emission onset, a direct band gap reaching from the visible range down to the near infrared (figure 7.1), long lifetimes of excited charge carriers, as well as ease of fabrication from simple salt precursor solutions, halide perovskites quickly found use in other applications. Closely related to solar cells, halide perovskite photodetectors make use of the unique optoelectronic properties and in particular can allow for colour sensitive detection based on the tuneable band gap via the aforementioned compositional engineering. The inverse process of light absorption and the conversion of a photon's energy into excited charge carriers in a semiconductor's conduction band is light emission by radiative recombination. This process allows the use of halide perovskites in perovskite light emitting devices (P-LEDs) and perovskite lasers (Veldhuis *et al* 2016). In order to obtain light emission with an even narrower emission peak, researchers successfully produced halide perovskite quantum dots (Wang *et al* 2017b). Another application that has been explored with halide perovskites is x-ray detection. Since OMH perovskites contain halides with high electron densities as well as Pb or Sn, the absorption coefficient of these crystals is sufficiently high in order to allow a high photon absorption coefficient and sensitivity for x-rays. The excited charge carriers far exceed the band gap energy of these perovskites and after relaxation to the band edges the increased conductivity can be detected by drift current measurements (Wei and Huang 2019).

7.2 Controversy around ferroelectricity in methylammonium lead iodide

In 2014, Frost *et al* proposed the concept that light-absorbing semiconductors which also exhibit spontaneous ferroelectric polarization might exhibit favourable electronic properties for photovoltaic applications and, that methylammonium lead iodide ($MAPbI_3$) might indeed be such a material (Frost *et al* 2014). They concluded that a semiconducting ferroelectric that spontaneously forms domains of alternating polarization could lead to reduced Shockley–Read–Hall (SRH) recombination. Yet, at that time, there was no experimental proof for ferroelectricity in halide perovskites and this model was mainly based on simulations and theoretical considerations. Later in 2014, Kutes *et al* reported the observation of ferroelectric domains in $MAPbI_3$ via piezoresponse force microscopy (PFM) but they did not observe ordered domains which typically form in classic ferroelectrics (Kutes *et al* 2014). In 2016 and 2017 two following reports by Hermes *et al* and Röhm *et al* revealed highly ordered domain structures in $MAPbI_3$, as shown in figures 7.2(a) and (b) (Hermes *et al* 2016, Röhm *et al* 2017).

Hermes *et al* assigned these domains to ferroelasticity since ferroelectric poling could not be readily observed in this material, while Röhm *et al* concluded that the domains are ferroelectric based on their alternating piezoresponse and their effect on charge carrier extraction in conductive atomic force microscopy (AFM) measurements. In the following years, multiple studies investigated the nature of these domains with seemingly contradictory results. Liu *et al* showed that the surface of domains in $MAPbI_3$ sometimes exhibit a chemical contrast of methylammonium concentration. They assigned this effect to ferroelastic twinning and concluded that this chemical segregation is the reason for a contrast in PFM measurements (Liu *et al* 2018). In particular, they pointed out that the use of signal amplification near the contact resonance frequency of the AFM cantilever makes the signal susceptible to misinterpretation. This study also observed the same twin domains in band-excited PFM, yet Liu *et al* pointed out that this could be due to local contact resonance variations that might be caused by ferroelastic strain. In a response, Schulz *et al* argued that these effects are not contradictory to ferroelectricity since all ferroelectrics are also ferroelastic due to their non-centrosymmetry and that a chemical contrast via screening ions at the surface of such domains is in fact to be expected (Schulz *et al* 2019). Nevertheless, both viewpoints were supported by valid arguments which kept the discussion open (Liu *et al* 2019).

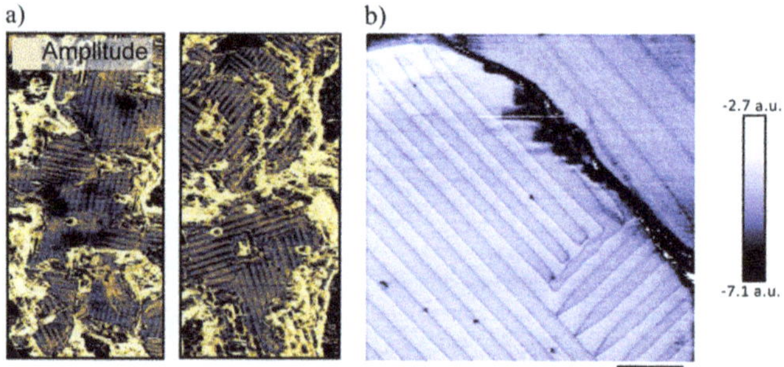

Figure 7.2. (a) PFM amplitude image of a $MAPbI_3$ sample by Weber *et al* assigned to ferroelastic twinning. (Adapted with permission from Hermes *et al* (2016), copyright © 2016 American Chemical Society.) (b) PFM inphase image of $MAPbI_3$ by Röhm *et al* claiming ferroelectric properties. (Scale bar 360 nm.) Adapted from Röhm *et al* (2017) with permission from the Royal Society of Chemistry.

One particular property of $MAPbI_3$ solar cells that was initially suspected to be caused by ferroelectricity is a hysteresis in current–density–voltage (JV) sweeps (Wei *et al* 2014). Yet, as it turned out, this hysteresis is mainly related to changes of ionic charge carrier accumulation or depletion at interfaces between the perovskite film and charge carrier transport or electrode layers (Weber *et al* 2018). Although ferroelectric domains and ferroelectric poling could influence these properties as well, the majority of the observed hysteresis effects are likely not caused by ferroelectric poling. Leguy *et al* argued that based on Monte Carlo simulations, the timescales of ferroelectric switching, if related to the alignment of MA^+ dipole molecules, would be on the order of 1 ms and hence faster than the experimentally observed effects in perovskite solar cells (Leguy *et al* 2015). This conclusion is in line with the observation of similar JV-hysteresis in metal halide perovskites that possess a centrosymmetric cubic structure and hence a paraelectric crystal phase at room temperature such as $FAPbI_3$ and $MAPbBr_3$.

The crystal structure and hence the space group of all halide perovskites is decisive for the possibility for, or exclusion of, ferroic properties. To date, some uncertainty remains regarding the allocation of space groups for some compounds. This is often based on limitations to the accuracy of diffraction methods since even tiny offsets in the atomic positions in the crystal lattice can make the difference between macroscopically polar and nonpolar materials (Richter *et al* 2018). For archetypical $MAPbI_3$ and similar metal halides, instability under irradiation and a significant amount of dynamic disorder at elevated temperatures further complicate these measurements (Poglitsch and Weber 1987). Breternitz *et al* recently showed that an offset of iodine atoms in the crystal lattice supports the allocation of the non-centrosymmetric and even polar tetragonal space group I4cm for $MAPbI_3$, which is a prerequisite for ferroelectricity (Breternitz *et al* 2019, 2020). At higher temperatures (>327 K), $MAPbI_3$ presents itself in the cubic and thereby paraelectric space group Pm3m (Poglitsch and Weber 1987). The phase transition temperature is very close to room temperature, which not only indicates a low energetical difference between the tetragonal and cubic crystal symmetry as pointed out by Breternitz *et al* but also requires carful temperature control during commonly employed x-ray diffraction (XRD) measurements on this material. Unless conducted under temperature-controlled conditions, these measurements can be distorted by local heat sources such as illumination or environmental influences, which could unintentionally lead to measurements close to or above the Curie temperature (T_C) and thereby skew the results.

One of the biggest concerns in the discussion about the nature of the previously reported domain structures in $MAPbI_3$ is the lack of saturating polarization (P)–electric field (E) ferroelectric hysteresis loops (PE loops) as one of the hallmarks for ferroelectricity. Whenever an electric field is applied to a $MAPbI_3$ sample, a current composed of free electronic charge carriers and mobile ions is driven through the crystal. Besides the thermally activated electrons and holes, any above band gap (>1.6 eV) photon illumination further increases the charge carrier density and thereby the sample's conductivity. The ionic currents are temperature dependent with relatively low activation energies of ~0.29 eV and ~0.9 eV for I^- and MA^+, respectively (Futscher *et al* 2019). The resulting overall conductivity leads to thermal damage to $MAPbI_3$ samples, even at moderate electric poling fields on the order of 5 V μm^{-1}. Figure 7.3(a) depicts a PE-curve of a perovskite thin film sample in a typical solar cell architecture. Even at small electric fields of 1.5 V μm^{-1}, a current occurs that quickly rises and does not reach saturation before thermal destruction occurs. This can be avoided by limiting the electric field to approximately 3.0 V μm^{-1}, but mainly capacitive behaviour with a significant leakage (ohmic/shunt) component can be observed in the resulting PE-curve. In contrast, classic insulating ferroelectrics such as $BiFeO_3$ retain a very small current up to ~20 V μm^{-1} before ferroelectric switching mediates a poling current (figure 7.3(b)).

In order to limit ionic contributions to the leakage current of $MAPbI_3$, Rakita *et al* cooled samples to 204 K and observed a saturation in the complex part of the dielectric response during AC poling experiments (Rakita *et al* 2017). Compared to ceramic ferroelectrics, in $MAPbI_3$ thermal degradation sets in at comparatively low temperatures of ~420 K, limiting the maximum current–density in poling experiments (Kim *et al* 2018a). Additionally, thin film samples in solar-cell architectures (figure 7.3(a)) often show strong [110] texture hence confining the *c*-axis of the tetragonal crystals to an in-plane orientation parallel to the substrate and orthogonal to the electric field between electrodes (Leonhard *et al* 2019, Wu *et al* 2018). Therefore, the magnitude of the minimal electric field that might induce quick ferroelectric switching (coercive field) is increased compared to experiments where the poling field aligns with the *c*-axis of the crystalline sample.

In order to minimize the electric field that is required to achieve ferroelectric switching, Röhm *et al* used a sample layout of interdigitated gold electrodes covered by a thin film of $MAPbI_3$ for ferroelectric poling experiments (figures 7.4(a) and (b)) (Röhm *et al* 2019).

In this configuration, the poling field between the gold electrodes lies in the same plane as the tetragonal *c*-axis of the [110] textured film. Since the current response of $MAPbI_3$ samples shows a diode-like onset behaviour (figure 7.3(a)), the minimization of the electric field maximum is key in order to prevent thermal damage to the

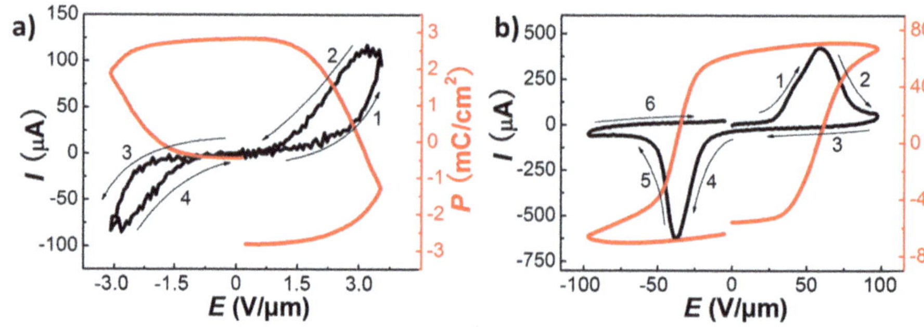

Figure 7.3. (a) Polarization curve of an ITO/PEDOT:PSS/$MAPbI_3$/Au thin film sample (10 Hz) that does show semiconducting and capacitive behaviour unlike classic insulating ferroelectric samples. (b) Reference measurement of a ferroelectric capacitor of Au/$BiFeO_3$/SrRuO (1000 Hz) showing clear saturation of the polarization curve. Adapted with permission from Fan *et al* (2015). Copyright © 2015 American Chemical Society.

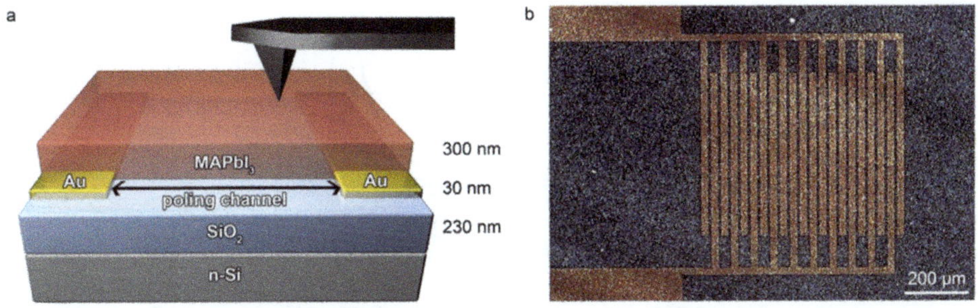

Figure 7.4. (a) Schematic of interdigitated gold electrodes atop an n-doped silicon substrate allowing the application of an in-plane poling field to the $MAPbI_3$ thin film and surface sampling via AFM before and after poling. The insulating SiO_2 layer prevents leakage currents through the substrate. (b) A light microscope image shows the $MAPbI_3$ thin film covering the electrode layout. Reproduced according to CC BY-SA 4.0, Röhm *et al* (2019).

sample due to high currents. Accordingly, application of a small electric field for an extended period of time is a promising strategy. In the past, this creeping poling was successfully used in ceramic PZT ferroelectrics in order to reduce the electric field required for ferroelectric poling (Zhukov *et al* 2010). The current characteristics of Röhm *et al*'s experiment are depicted in figure 7.5. The pristine sample shows a symmetrical and reversible current (I)–E curve, similar to a double-Schottky diode. At a characteristic onset field (E_{on}) of ± 1.6 V μm^{-1}, the electric current increases exponentially, indicative of the Schottky contact at the semiconductor/metal interface. The pronounced hysteresis is likely related to accumulation of MA^+ or I^- ions at the interfaces. In contrast, during creeping poling of the $MAPbI_3$ film at an electric field of 2 V μm^{-1} for 10 min between the interdigitated gold electrodes, the current stays below 1 µA yet results in a dramatic change of the subsequent IE-sweep characteristic (figure 7.5(b)). The poled sample possesses a diode behaviour with a reduced onset voltage of ~0.5 V μm^{-1} and current blocking at negative bias. The poling process is fully reversible via creeping poling for several minutes with opposite polarity, resulting in mirrored diode behaviour (figure 7.5(c)). The creeping poling currents over time (figures 7.5(d) and (e)) reveal slow poling processes over several minutes without any signs for burnout or thermal damage to the sample.

Nevertheless, based on the aforementioned properties the contribution of ferroelectric poling currents, ionic currents and even thermal degradation of the sample is hard to disentangle in macroscopic electrical measurements in $MAPbI_3$ and similar metal halide perovskites hence calling for microscopic investigation of the process.

7.3 Challenges for measuring ferroelectric nanostructures in methylammonium lead iodide

Unlike most functional ceramics, OMH perovskites are rather soft materials that are susceptible to damage from mechanical forces, changes in temperature and electric or ionic currents induced by measurement instruments (Li *et al* 2019, Feng 2014, Sun *et al* 2017). Additionally, extensive measurements on thin films of hygroscopic

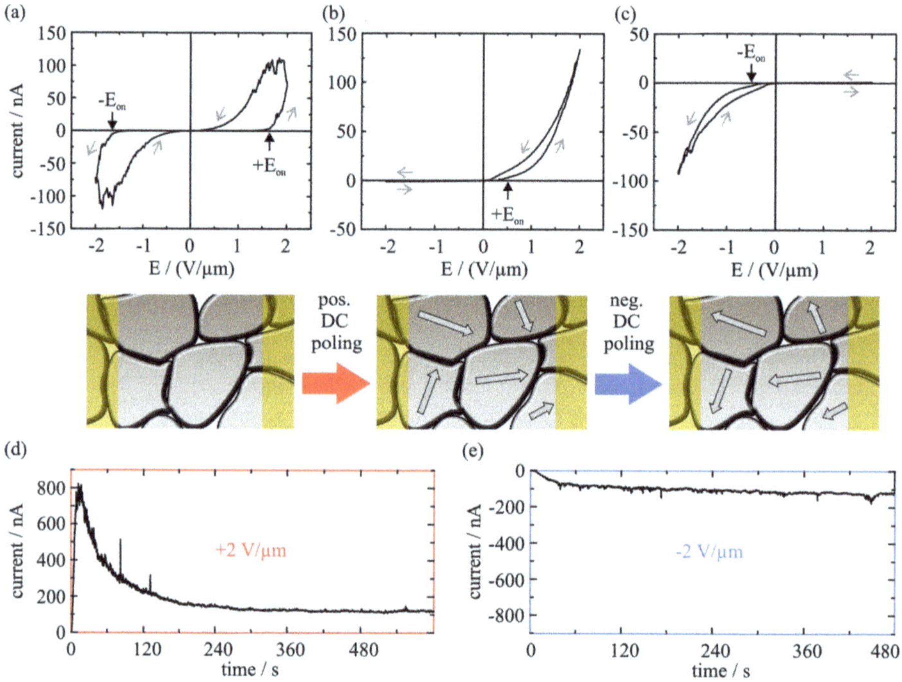

Figure 7.5. (a) Electric field sweeps between interdigitated gold electrodes result in a symmetrical double-Schottky diode behaviour of the Au/$MAPbI_3$/Au sample. (b) After application of an electric field of 2 V μm^{-1} for 10 min (creeping poling), the electrical conductivity of the device becomes directional, indicating a stable polarization of the $MAPbI_3$ film. (c) The directional behaviour can be inversed by application of an inverted poling field for several minutes. (d), (e) The device current gives insight into the changes of sample conductivity during creeping poling. After initial poling of the pristine sample, the sample current reaches a stable magnitude of ±120 nA for both poling directions. Reproduced according to CC BY-SA 4.0, Röhm (2019).

materials such as $MAPbI_3$ must be conducted under dry gas atmospheres in order to avoid decomposition.

Exertion of mechanical forces or generation of mechanical stress in the sample are a pivotal part of many ferroic characterization techniques such as dilatometry, piezoresponse measurements and mechanical profilometry. Even contact mode scanning probe microscopy such as PFM and conductive AFM, which only exhibit very small forces can lead to abrasion of OMH film surfaces. This type of measurement damage is most likely to occur when the uneven topography of a polycrystalline layer is scanned at a fast rate (Hieulle *et al* 2017, Leonhard *et al* 2019). To some extent, sample damage with scanning probe measurements can be minimized by using softer cantilevers, but this approach renders the measurements more susceptible to artifacts such as electrostatic noise (Collins *et al* 2019). The mapping of domains in $MAPbI_3$ via PFM is not only challenging because of these mechanical limitations and the relatively low durability of the samples, but since this perovskite is both semiconducting for electronic charge carriers and an ionic conductor (MA^+, I^-), these contributions will also affect any kind of measurement that modulates an electric field in the sample.

In PFM, typical excitation voltages of 1–2 V are applied to the AFM tip, which translate to electric fields of 1–10 V μm^{-1} at several kHz through the polycrystalline thin film, depending on its thickness. To what extent the resulting movement of the AFM tip is caused by a true piezoresponse of $MAPbI_3$ or by electrostatic or mechanical artifacts has fired up a lively scientific discussion (Schulz *et al* 2019, Liu *et al* 2019, Colsmann *et al* 2020, Gomez *et al* 2020).

While PFM does not directly measure polarization of a potential ferroelectric material but rather its local piezoresponse, it offers the possibility to directly map piezoresponsive domains that are typically on the order of 100 nm in size in a nondestructive manner. By using a device layout with a $MAPbI_3$ film atop interdigitated gold electrodes on an insulated silicon substrate as shown in figure 7.4, PFM measurements before and after each poling experiment can reveal changes in the sample (Röhm *et al* 2019).

If the applied electric field affects the domains (ferroelectric poling), this can be observed directly without the need for analysis of the macroscopic poling current. Particularly on semiconducting samples, this approach does not require the distinction between parasitic conductivity and a poling current, which is used for classic hysteresis curve analysis. Figure 7.6 shows lateral PFM measurements of two different positions on a $MAPbI_3$ film in their pristine state and after creeping poling at 4.5 V μm^{-1} for 11 min.

Some clear changes in the domain patterns become apparent after poling. Not only do positions of the domain walls shift due to electric poling, but also the ratio of both types of domains within grains is altered (figures 7.6(d) and (e)). Generally, it is challenging to deduct the polarization direction of ferroelectric domains via PFM, but the combination of a well-known electrical poling field, the changes of domains before and after poling and the assumption of 90° domain walls based on the crystal structure allows for general conclusions on the polarization direction in this experiment (arrows in figures 7.6(d) and (e)). Since neighbouring domains show a clear contrast in lateral PFM amplitude, 180° domains can be ruled out (Röhm *et al* 2017).

Besides PFM, scanning electron microscopy (SEM) or scanning transmission electron microscopy (STEM) are techniques that can reveal ferroelectric domains and even crystal lattice resolution of the atomic and molecular arrangement within crystals, as shown in figure 7.7. However, they require carefully chosen measurement parameters in order to avoid damage to the sample especially in metal halide perovskites. The most common issue is direct beam damage where the beam of accelerated electrons degrades the perovskite (Rothmann *et al* 2017a). Both the acceleration voltage (energy per electron) and the beam current (rate of electron incidence) have to be within a certain range in order to obtain a sufficient measurement signal from the detectors as well as to correlate the signal with energy input into the specimen, as explained in figure 7.8. In particular, not only are OMH perovskites susceptible to decomposition of the organic compound due to this energy influx and resulting heating and ionization, but also inorganic metal halides such as $CsPbI_3$ can be damaged, even though they show more resilience (Xiao *et al* 2015).

Resolving crystal orientation with sub-grain resolution in samples with micrometer sized grains is often required in order to link measurement results from other

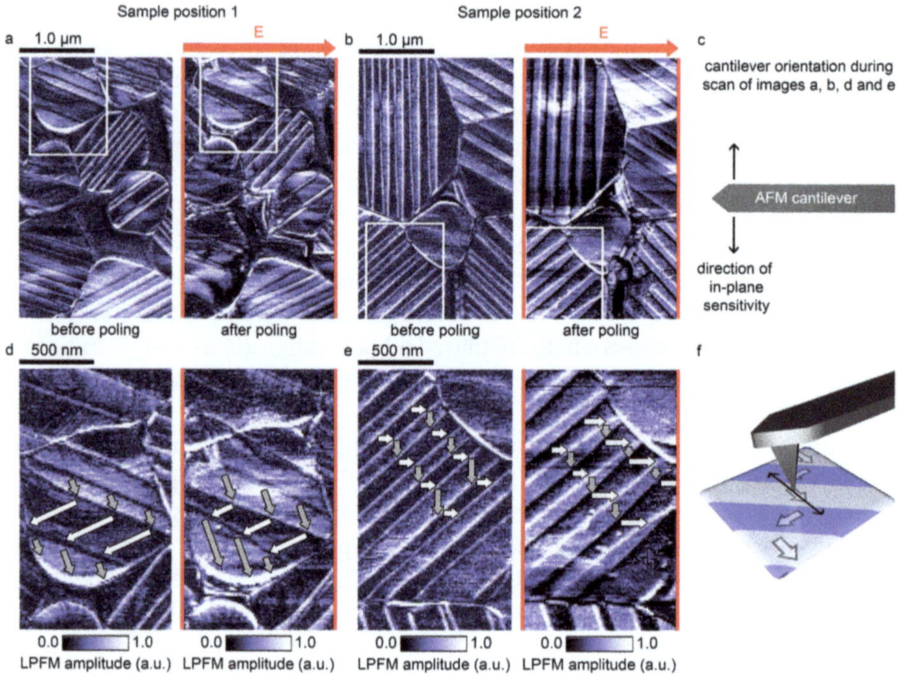

Figure 7.6. (a) Lateral PFM images of $MAPbI_3$ grains atop a silicon substrate (see figure 7.4(a)). The domain pattern of the pristine sample (left) changes drastically after application of a DC (direct current) poling field (red markings, 4.5 V μm^{-1}, 11 min). (b) At a different sample position widening of domains with a low lateral piezoresponse (dark) can be observed after poling while domains with a high lateral piezoresponse (light blue) shrink in size. By combining these results with the orientation of the electric poling field (red arrows) and the direction of the AFM cantilever during the measurement (c), a probable configuration of domain polarization can be deduced (d)–(f). Reproduced according to CC BY-SA 4.0, Röhm *et al* (2019).

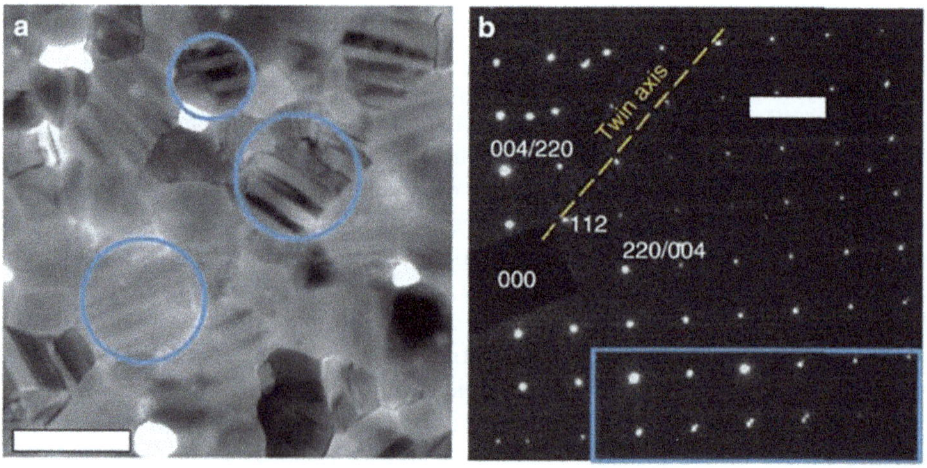

Figure 7.7. (a) Bright-field TEM image of $MAPbI_3$ grains show a clear contrast of neighbouring domains (scale bar 500 nm). (b) Diffraction pattern taken from a grain exhibiting stripe contrast (scale bar 2 nm). Reproduced according to CC BY-SA 4.0, Rothmann *et al* (2017b).

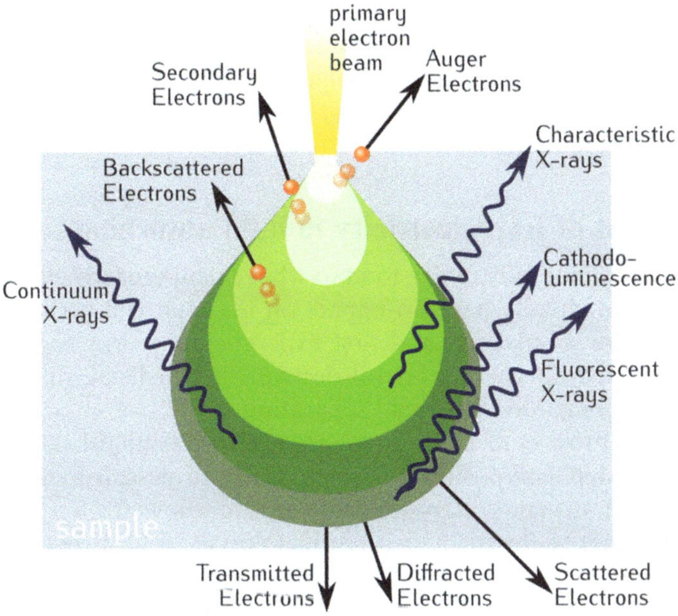

Figure 7.8. Overview for interactions of an electron beam with a sample. Depending on the type of interaction, the measurement can introduce significant degradation to the sample, limiting the measurement duration and other measurement parameters when probing OMH perovskites. Image credit: Wikipedia user Ponor, distributed under CC BY-SA 4.0.

microscopy techniques such as PFM to the crystal lattice. Electron backscatter diffraction (EBSD), which is based on SEM, is one of the few microscopy techniques that can provide this information in a nondestructive manner. In EBSD, the electron beam of an SEM is probing the sample surface at an angle of 20° and the resulting Bragg-diffraction pattern is measured on a detector screen. This interference pattern is then matched to a crystallographic orientation based on the appropriate space group. This software-based matching requires a sufficient signal amplitude in the detector and a signal contrast that can be matched to simulated patterns. In OMH perovskites, the duration of electron beam irradiation for each sample position is severely limited by the resulting beam damage, and thus limiting the amount of backscattered electrons that reach the detector. One promising approach has been to reduce sample degradation of $MAPbI_3$ during EBSD measurements by using low-vacuum chamber pressures inside the SEM with up to 1 mbar of partial water vapour pressure which reduces charge accumulation on each sample position. In this way, grain orientation mapping with sub-grain resolution of $MAPbI_3$ thin films was achieved for the first time in 2019 (Leonhard *et al* 2019). This allowed for correlation of grain orientation, ferroelectric domain patterns mapped via PFM as well as contact potential measurements via Kelvin probe force microscopy (KPFM). Soon after, using ultrafast EBSD detectors, the resolution of EBSD measurements was further improved, even revealing intra-grain strain in $MAPbI_3$ thin films (Jariwala *et al* 2019). These results show that many established microscale characterization techniques that have been used on ceramic ferroelectrics in the past can also be adapted to OMH perovskites, but the mechanical and chemical instability of these compounds often require adjusted measurement parameters and sometimes even technological improvements.

7.4 Implications of ferroelectricity in light absorbing semiconductors

At this time, it is still not fully understood to what extent the ferroelectric domains in $MAPbI_3$ solar cells actually have an impact on the solar cell performance. On the one hand, there are simulations and first experimental works suggesting that the formation of ordered alternating lateral domains is beneficial, mainly in order to reduce SRH recombination via spatial separation of electrons and holes within the perovskite grains (Frost *et al* 2014, Rossi *et al* 2018, Leonhard *et al* 2021). On the other hand, perovskite compositions that are likely to possess a cubic and thereby paraelectric crystal structure at room temperature allow for very high solar cell efficiencies very close to the theoretical limits (Yoo *et al* 2021, Hui *et al* 2021).

One of the biggest challenges for separating the various crystalline, ionic and electronic effects in OMH perovskites is the lack of comparability between different perovskite compositions. There is no way of making a perovskite solar cell with and without ferroelectric properties without changing its crystal structure and often even the thin film morphology. One seemingly obvious approach is to heat a ferroelectric perovskite solar cell beyond T_C and compare the device properties. But even though for example $MAPbI_3$ solar cells perform worse when heated to the cubic phase, ionic effects and thereby defect densities, shifts of the conduction and valence band energies as well as strain within the thin film could be reasonable explanations besides direct influence from ferroelectric domains (Schwenzer *et al* 2018, Zhu *et al* 2019). Additionally, both tetragonal and cubic phases can coexist in thin film samples of $MAPbI_3$ at room temperature and up to 330 K (Whitfield *et al* 2016, Kim *et al* 2018b, 2018c).

Unlike real experiments, simulations allow comparison of perovskites with identical properties besides crystal polarization and can thereby inform a deeper understanding of charge carrier dynamics. In order to yield meaningful results, the underlying assumptions and parameters must sufficiently reflect the real system. Rossi *et al* used finite-element discretization of drift-diffusion equations to simulate domain patters that were previously observed via PFM measurements on $MAPbI_3$ thin films (Rossi *et al* 2018). Figure 7.9 shows three different variations of ferroelectric domains, electron and hole densities along laterally cut planes, and the respective spatial SRH recombination rates. These simulations showed that specific ferroelectric domain configurations could affect recombination losses, particularly in thin films with high defect mediated trap-densities. Generally, the formation of lateral 90° domains with charged domain walls was predicted to have the most beneficial impact on device performance while uncharged domain walls are less likely to affect the solar cell. Vertical polarization components (perpendicular to the planar electrodes) do not affect charge carrier recombination rates but they

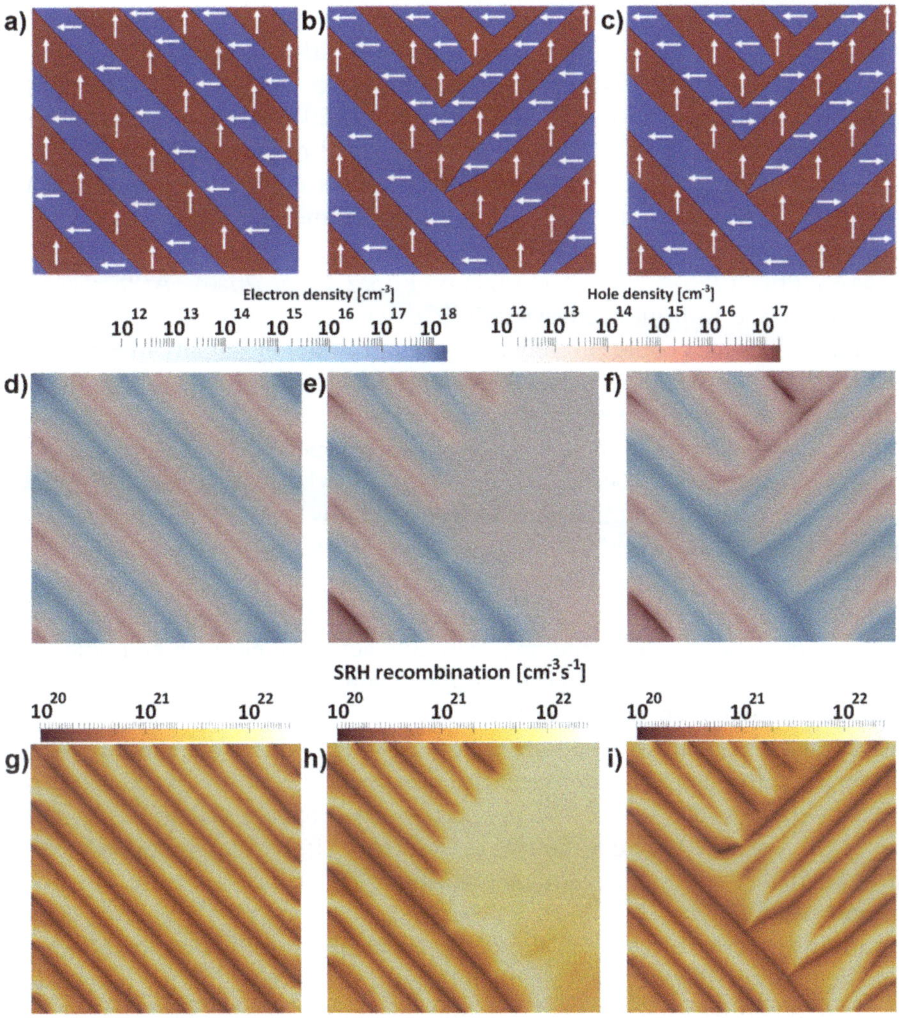

Figure 7.9. (a)–(c) Various models of lateral ferroelectric domain polarization patterns based on PFM measurements of $MAPbI_3$ thin films. (d)–(f) Drift-diffusion simulations of charge carrier densities during illumination of the modelled solar cell reveal separate areas of higher electron- or hole-densities in lateral cut planes of the crystal grain even for a moderate polarization of 0.2 μC cm^{-2}. (g)–(i) The resulting SRH recombination rate is significantly reduced in areas of the lowest overlap between electron- and hole-densities. Adapted with permission from Rossi *et al* (2018).

might affect ionic charge carrier densities and thereby the surface potentials at the interfaces towards neighbouring charge transport layers or electrodes. This effect could be utilized by applying vertical poling to solar cells in order to improve directional charge carrier transport and charge carrier extraction in unoptimized solar cell stacks as was previously demonstrated on the ferroelectric oxides Bi_2FeCrO_6 and KNN-BNNO (Ba and Ni doped $(K,Na)NbO_3$) (Nechache *et al* 2015, Bai *et al* 2018).

Although simulations have shown that some domain configurations of a semiconducting ferroelectric light absorbing material could reduce SRH-recombination losses, generally, ferroelectricity could also have detrimental effects on perovskite solar cells (Frost *et al* 2014, Rossi *et al* 2018). In particular, perovskite compositions that have a T_C within the operational range of solar cells will likely go through this transition when the solar cell heats up or cools down due to sun illumination in the morning, sunset in the evening and varying weather conditions. This is the case for many OMH perovskites, e.g. $MAPbI_3$ (T_C ~ 55 °C) and $FA_{0.1}MA_{0.9}PbI_3$ (T_C ~ 25 °C) (Francisco-López *et al* 2020). A ferroelectric to paraelectric phase transition always coincides with the release of screening charges at domain walls which could severely impact compositional stability in a material like $MAPbI_3$ that possesses a low activation energy for its A- and X-site ions (MA^+, I^-). Repeated trapping and release of these ions at domain walls could mediate crystal decomposition or severely change its electronic properties (Colsmann *et al* 2019). A similar process might occur at domain walls whenever a ferroelectric OMH perovskite is exposed to illumination. At this point it is still unclear whether screening charges at domain walls mainly comprise ionic or electronic charge carriers but, it seems plausible that in dark conditions these screening charges are predominantly ionic, since previously photogenerated free electronic charges (electrons and holes) are likely to recombine and the remaining thermally activated electron and hole densities are much lower. Therefore, a $MAPbI_3$ sample might release ionic charge carriers from domain walls, once an abundance of electrons and holes are generated under illumination that can hence contribute to domain polarization screening. Conversely, when illumination is removed, ionic charges are more likely to get trapped at domain walls until the polarization is screened. These processes might contribute to the slow time constants and capacitive effects that are often observed in solar simulator or external quantum efficiency (EQE) measurements in perovskite solar cells (Ravishankar *et al* 2018, Röhm 2019).

Additionally, crossing the tetragonal to cubic phase transition temperature will lead to fluctuations of stress and hence will result in strain within the perovskite thin film. These repetitive changes in strain could not only lead to defect formation and changes of the interfaces between perovskite grains and surrounding charge transport layers in a solar cell, but this will likely also impact grain morphology and properties of grain boundaries over time. These effects might even be one of the crucial limiting factors to the long-term stability of some OMH-perovskite compositions. Besides device stability, tensile strain in OMH solar cells can also directly or indirectly change electronic properties and thereby impact solar cell performance (Zhu *et al* 2019). Besides temperature effects around the phase transition in a light-absorbing perovskite thin film, electric fields might play a role when the material is ferroelectric. Both during solar cell operation and device testing, electric fields on the order of 1–10 V μm^{-1} can occur within the perovskite film depending on the layer thickness, illumination intensity and external bias. As discussed in the previous sections of this chapter, ferroelectric domains in $MAPbI_3$ are affected by electric fields of this magnitude if they are applied for several minutes. Whether or not ferroelectric poling occurs depends on strain constraints (clamping, pinning, tetragonality), the alignment of the electric field relative to the crystal polarization, and the temperature. This potential for creeping poling in ferroelectric OMH-perovskite solar cells seems very likely to mediate changes to electronic properties and to impact device stability over time.

7.5 What is next for metal halide perovskites?

Metal halide perovskites are currently at the crossroads where several pioneering companies have started to commercialize their products, but it is still unclear which markets will become most feasible. Will the tandem approach bring sufficient improvements to the efficiency of solar modules by harvesting high-energy photons with a perovskite solar cell on top of a regular silicon solar cell? Can the lifetime of the metal halide perovskite devices match the excellent lifetimes of silicon solar cells that already exceed 30 years? Or will perovskite solar cells by themselves cross the barrier of becoming competitive with silicon solar cells by outcompeting them in terms of production cost? If these scenarios do not come to fruition, it will most likely be due to limitations of stability of the perovskite crystals or the currently still unavoidable incorporation of the heavy metal lead.

Intrinsic material properties like ferroelectricity, ferroelasticity and phase transitions will play a pivotal role not only in understanding the photophysics and charge carrier dynamics in perovskite solar cells but also in improving the currently used perovskite compositions in terms of stability and power conversion efficiency. Development of lead-free alternative materials with similar properties will certainly require a good understanding of all the physical processes in the crystal under illumination. The same holds true for other applications of this fascinating material class such as perovskite-LEDs, perovskite photodetectors or perovskite electronics. Whether ferroelectricity turns out to be beneficial or detrimental to perovskite charge carrier generation, charge carrier transport and recombination dynamics, it certainly will be a key parameter to optimize. Most state-of-the-art OMH perovskite compositions comprise tetragonal to cubic phase transitions within the solar cells' operational temperature range, but these devices are most often only optimized for standard testing conditions (e.g. 25 °C) and it is still unclear whether these properties are sustained at higher or lower operation conditions that can entail a different crystal phase. Understanding and controlling these phase transitions will certainly play an important role in improving the long-term stability and average power conversion efficiencies of perovskite solar cells outside today's research labs.

References

Bai Y, Vats G, Seidel J, Jantunen H and Juuti J 2018 Boosting photovoltaic output of ferroelectric ceramics by optoelectric control of domains *Adv. Mater.* **30** 1803821

Breternitz J, Lehmann F, Barnett S A, Nowell H and Schorr S 2019 Role of the iodide–methylammonium interaction in the ferroelectricity of $CH_3NH_3PbI_3$ *Angew. Chem. Int. Ed.* **59** 424–8

Breternitz J, Tovar M and Schorr S 2020 Twinning in $MAPbI_3$ at room temperature uncovered through Laue neutron diffraction *Sci. Rep.* **10** 16613

Chao L, Wang Z, Xia Y, Chen Y and Huang W 2018 Recent progress on low dimensional perovskite solar cells *J. Energy Chem.* **27** 1091–100

Chen J, He D and Park N-G 2021 Methodologies for >30% efficient perovskite solar cells via enhancement of voltage and fill factor *Sol. RRL* **6** 2100767

Collins L, Liu Y, Ovchinnikova O S and Proksch R 2019 Quantitative electromechanical atomic force microscopy *ACS Nano* **13** 8055–66

Colsmann A and Röhm H 2019 Ferroelectricity and stability measurements in perovskite solar cells *J. Phys.: Energy* **2** 011003

Colsmann A, Leonhard T, Schulz A D and Röhm H 2020 Comment on 'ferroelectricity-free lead halide perovskites' by A. Gómez et al. *Energy Environ. Sci.* **13** 1888–91

Fan Z, Xiao J, Sun K, Chen L, Hu Y, Ouyang J, Ong K P, Zeng K and Wang J 2015 Ferroelectricity of $CH_3NH_3PbI_3$ perovskite *J. Phys. Chem. Lett.* **6** 1155–61

Feng J 2014 Mechanical properties of hybrid organic-inorganic $CH_3NH_3BX_3$ (B = Sn, Pb; X = Br, I) perovskites for solar cell absorbers *APL Mater.* **2** 081801

Francisco-López A, Charles B, Alonso M I, Garriga M, Campoy-Quiles M, Weller M T and Goñi A R 2020 Phase diagram of methylammonium/formamidinium lead lodide perovskite solid solutions from temperature-dependent photoluminescence and raman spectroscopies *J. Phys. Chem.* C **124** 3448–58

Frost J M and Walsh A 2016 What is moving in hybrid halide perovskite solar cells? *Acc. Chem. Res.* **49** 528–35

Frost J M, Butler K T, Brivio F, Hendon C H, van Schilfgaarde M and Walsh A 2014 Atomistic origins of high-performance in hybrid halide perovskite solar cells *Nano Lett.* **14** 2584–90

Futscher M H, Lee J M, McGovern L, Muscarella L A, Wang T, Haider M I, Fakharuddin A, Schmidt-Mende L and Ehrler B 2019 Quantification of ion migration in $CH_3NH_3Pb_{I3}$ perovskite solar cells by transient capacitance measurements *Mater. Horiz.* **6** 1497–503

Gomez A, Wang Q, Goñi A R, Campoy-Quiles M and Abate A 2020 Reply to the "Comment on the publication 'Ferroelectricity-free lead halide perovskites' by Gomez *et al.*" by Colsmann *et al. Energy Environ. Sci.* **13** 1892–5

Green M A, Ho-Baillie A and Snaith H J 2014 The emergence of perovskite solar cells *Nat. Photonics* **8** 506–14

Hermes I M *et al* 2016 Ferroelastic fingerprints in methylammonium lead iodide perovskite *J. Phys. Chem.* C **120** 5724–31

Hieulle J, Stecker C, Ohmann R, Ono L K and Qi Y 2017 Scanning probe microscopy applied to organic–inorganic halide perovskite materials and solar cells *Small Methods* **2** 1700295

Hsieh C-M, Yu Y-L, Chen C-P and Chuang S-C 2017 Effects of the additives n-propylammonium or n-butylammonium iodide on the performance of perovskite solar cells *RSC Adv.* **7** 55986–92

Hui W *et al* 2021 Stabilizing black-phase formamidinium perovskite formation at room temperature and high humidity *Science* **371** 1359–64

Jariwala S, Sun H, Adhyaksa G W P, Lof A, Muscarella L A, Ehrler B, Garnett E C and Ginger D S 2019 Local crystal misorientation influences non-radiative recombination in halide perovskites *Joule* **3** 3048–60

Kim H-S *et al* 2012 Lead iodide perovskite sensitized all-solid-state submicron thin film mesoscopic solar cell with efficiency exceeding 9% *Sci. Rep.* **2** 591

Kim T W, Shibayama N, Cojocaru L, Uchida S, Kondo T and Segawa H 2018a Real-time *in situ* observation of microstructural change in organometal halide perovskite induced by thermal degradation *Adv. Funct. Mater.* **28** 1804039

Kim T W *et al* 2018b Self-organized superlattice and phase coexistence inside thin film organometal halide perovskite *Adv. Mater.* **30** 1705230

Kim T W, Matsushita T, Uchida S, Kondo T and Segawa H 2018c Quantitative fraction analysis of coexisting phases in a polycrystalline $CH_3NH_3PbI_3$ perovskite *Appl. Phys. Express* **11** 101401

Kojima A, Teshima K, Miyasaka T and Shirai Y 2006 Novel photoelectrochemical cell with mesoscopic electrodes sensitized by lead-halide compounds (2) *Meet. Abstr.* **MA2006-02** 397

Kutes Y, Ye L, Zhou Y, Pang S, Huey B D and Padture N P 2014 Direct observation of ferroelectric domains in solution-processed $CH_3NH_3PbI_3$ perovskite thin films *J. Phys. Chem. Lett.* **5** 3335–9

Lee M M, Teuscher J, Miyasaka T, Murakami T N and Snaith H J 2012 Efficient hybrid solar cells based on meso-superstructured organometal halide perovskites *Science* **338** 643–47

Leguy A M A *et al* 2015 The dynamics of methylammonium ions in hybrid organic–inorganic perovskite solar cells *Nat. Commun.* **6** 7124

Leonhard T, Schulz A D, Röhm H, Wagner S, Altermann F J, Rheinheimer W, Hoffmann M J and Colsmann A 2019 Probing the microstructure of methylammonium lead iodide perovskite solar cells *Energy Technol.* **7** 1800989

Leonhard T, Röhm H, Altermann F J, Hoffmann M J and Colsmann A 2021 Evolution of ferroelectric domains in methylammonium lead iodide and correlation with the performance of perovskite solar cells *J. Mater. Chem.* A **9** 21845–58

Li X, Fu S, Liu S, Wu Y, Zhang W, Song W and Fang J 2019 Suppressing the ions-induced degradation for operationally stable perovskite solar cells *Nano Energy* **64** 103962

Liu Y *et al* 2018 Chemical nature of ferroelastic twin domains in $CH_3NH_3PbI_3$ perovskite *Nat. Mater.* **17** 1013–19

Liu Y *et al* 2019 Reply to: On the ferroelectricity of $CH_3NH_3PbI_3$ perovskites *Nat. Mater.* **18** 1051–3

Nechache R, Harnagea C, Li S, Cardenas L, Huang W, Chakrabartty J and Rosei F 2015 Bandgap tuning of multiferroic oxide solar cells *Nat. Photon.* **9** 61–7

Ono L K, Liu S F and Qi Y 2019 Reducing detrimental defects for high-performance metal halide perovskite solar cells *Angew. Chem. Int. Ed.* **59** 6676–98

Pecchia A, Gentilini D, Rossi D, Auf der Maur M and Di Carlo A 2015 Role of ferroelectric nanodomains in the transport properties of perovskite solar cells *Nano Lett.* **16** 988–92

Poglitsch A and Weber D 1987 Dynamic disorder in methylammoniumtrihalogenoplumbates (II) observed by millimeter-wave spectroscopy *J. Chem. Phys.* **87** 6373

Rakita Y, Bar-Elli O, Meirzadeh E, Kaslasi H, Peleg Y, Hodes G, Lubomirsky I, Oron D, Ehre D and Cahen D 2017 Tetragonal $CH_3NH_3PbI_3$ is ferroelectric *Proc. Natl Acad. Sci.* **114** E5504–12

Ravishankar S, Aranda C, Boix P, Anta J A, Bisquert J and Garcia-Belmonte G 2018 Effects of frequency dependence of the external quantum efficiency of perovskite solar cells *J. Phys. Chem. Lett.* **9** 3099–104

Richter C, Zschornak M, Novikov D, Mehner E, Nentwich M, Hanzig J, Gorfman S and Meyer D C 2018 Picometer polar atomic displacements in strontium titanate determined by resonant X-ray diffraction *Nat. Commun.* **9** 178

Röhm H, Leonhard T, Hoffmann M J and Colsmann A 2017 Ferroelectric domains in methylammonium lead iodide perovskite thin-films *Energy Environ. Sci.* **10** 950–5

Röhm H, Leonhard T, Hoffmann M J and Colsmann A 2019 Ferroelectric poling of methylammonium lead iodide thin films *Adv. Funct. Mater.* **30** 1908657

Röhm H 2019 Ferroelektrizität in Methylammoniumbleiiodid-Solarzellen *PhD Thesis* Karlsruhe Institute of Technology, Karlsruhe

Rossi D, Pecchia A, Auf der Maur M, Leonhard T, Röhm H, Hoffmann M J, Colsmann A and Di Carlo A 2018 On the importance of ferroelectric domains for the performance of perovskite solar cells *Nano Energy* **48** 20–6

Rothmann M U, Li W, Etheridge J and Cheng Y-B 2017a Microstructural characterisations of perovskite solar cells—from grains to interfaces: techniques, features, and challenges *Adv. Energy Mater.* **7** 1700912

Rothmann M U, Li W, Zhu Y, Bach U, Spiccia L, Etheridge J and Cheng Y-B 2017b Direct observation of intrinsic twin domains in tetragonal $CH_3NH_3PbI_3$ *Nat. Commun.* **8** 14547

Saliba M *et al* 2016 Cesium-containing triple cation perovskite solar cells: improved stability, reproducibility and high efficiency *Energy Environ. Sci.* **9** 1989–97

Schulz A D, Röhm H, Leonhard T, Wagner S, Hoffmann M J and Colsmann A 2019 On the ferroelectricity of $CH_3NH_3PbI_3$ perovskites *Nat. Mater.* **18** 1050

Schwenzer J, Rakocevic L, Gehlhaar R, Abzieher T, Gharibzadeh S, Moghadamzadeh S, Quintilla A, Richards B S, Lemmer U and Paetzold U 2018 Temperature variation-induced performance decline of perovskite solar cells *ACS Appl. Mater. Interfaces* **10** 16390–9

Sun S, Isikgor F H, Deng Z, Wei F, Kieslich G, Bristowe P D, Ouyang J and Cheetham A K 2017 Factors influencing the mechanical properties of formamidinium lead halides and related hybrid perovskites *ChemSusChem* **10** 3740–5

Tai Q, Tang K-C and Yan F 2019 Recent progress of inorganic perovskite solar cells *Energy Environ. Sci.* **12** 2375–405

Tao S, Schmidt I, Brocks G, Jiang J, Tranca I, Meerholz K and Olthof S 2019 Absolute energy level positions in tin- and lead-based halide perovskites *Nat. Commun.* **10** 2560

Veldhuis S A, Boix P P, Yantara N, Li M, Sum T C, Mathews N and Mhaisalkar S G 2016 Perovskite materials for light-emitting diodes and lasers *Adv. Mater.* **28** 6804–34

Wang Z, Lin Q, Chmiel F P, Sakai N, Herz L M and Snaith H J 2017a Efficient ambient-air-stable solar cells with 2D–3D heterostructured butylammonium-caesium-formamidinium lead halide perovskites *Nat. Energy* **2** 17135

Wang H-C, Bao Z, Tsai H-Y, Tang A-C and Liu R-S 2017b Perovskite quantum dots and their application in light-emitting diodes *Small* **14** 1702433

Wang F, Bai S, Tress W, Hagfeldt A and Gao F 2018 Defects engineering for high-performance perovskite solar cells *npj Flex. Electron.* **2** 22

Weber S A L, Hermes I M, Turren-Cruz S-H, Gort C, Bergmann V W, Gilson L, Hagfeldt A, Graetzel M, Tress W and Berger R 2018 How the formation of interfacial charge causes hysteresis in perovskite solar cells *Energy Environ. Sci.* **11** 2404–13

Wei H and Huang J 2019 Halide lead perovskites for ionizing radiation detection *Nat. Commun.* **10** 1066

Wei J, Zhao Y, Li H, Li G, Pan J, Xu D, Zhao Q and Yu D 2014 Hysteresis analysis based on the ferroelectric effect in hybrid perovskite solar cells *J. Phys. Chem. Lett.* **5** 3937–45

Whitfield P S, Herron N, Guise W E, Page K, Cheng Y O, Milas I and Crawford M K 2016 Structures, phase transitions and tricritical behavior of the hybrid perovskite methyl ammonium lead iodide *Sci. Rep.* **6** 35685

Wilson J N, Frost J M, Wallace S K and Walsh A 2019 Dielectric and ferroic properties of metal halide perovskites *APL Mater.* **7** 010901

Wu H-T, Chen Y-F, Shih C-F, Leu C-C and Wu S-H 2018 Memory properties of (110) preferring oriented $CH_3NH_3PbI_3$ perovskite film prepared using PbS-buffered three-step growth method *Thin Solid Films* **660** 320–7

Xiao C *et al* 2015 Mechanisms of electron-beam-induced damage in perovskite thin films revealed by cathodoluminescence spectroscopy *J. Phys. Chem.* C **119** 26904–11

Yoo J J *et al* 2021 Efficient perovskite solar cells via improved carrier management *Nature* **590** 587–93

Zhu C *et al* 2019 Strain engineering in perovskite solar cells and its impacts on carrier dynamics *Nat. Commun.* **10** 815

Zhukov S, Fedosov S, Glaum J, Granzow T, Genenko Y A and von Seggern H 2010 Effect of bipolar electric fatigue on polarization switching in lead-zirconate-titanate ceramics *J. Appl. Phys.* **108** 014105

www.ingramcontent.com/pod-product-compliance
Ingram Content Group UK Ltd.
Pitfield, Milton Keynes, MK11 3LW, UK
UKHW051850100926
8556IPUK00005B/550